Fail Fast – Fail Safe

It's time to think, learn and work differently. Harness the power of failure to fearlessly innovate and achieve amazing results. Stop crashing and burning. Discover safe spots to land. Achieve more by learning how to Fail Fast – Fail Safe.

Paul Crosby, PMP®
Bob Prentiss, CBAP®

Fail Fast – Fail Safe
By Paul Crosby, PMP® and Bob Prentiss, CBAP®
© 2018 The League of Analysts Inc

All rights reserved. No part of this book may be reproduced or transmitted, in any form or by any means, electronic or mechanical, including photocopying, recording or by any other information storage and retrieval system, without the express written permission of the author except for brief quotations in review.

ISBN: 978-0-9975408-3-3

First Edition

Printed in the United States of America

This book is dedicated to every person who refuses to accept the status quo, accepts that we can always do better, and is willing to challenge and educate those who lack the understanding of what we try to do and the value it brings.

To all the amazing supporters of this book that have guided us both so much throughout this process.

We are humbled and grateful.

Table of Contents

Introducing Failure 7
Learning from Failure 49
Curiosity and Failure 127
Challenging Your Organization to Work with Failure 141
Fail Fast Model: Part 1 - Initiate 173
Fail Fast Model: Part 2 - Experiment 209
Fail Fast Model: Part 3 - Realization & Walking
Through the Fail Fast Model 245
Fail Safe Model 269
The Fear Model 291
Agile & Failure 315
Putting Fun in Failure 343
Think, Learn, Work Differently 357
About the Authors 382

Chapter 1

Introducing Failure

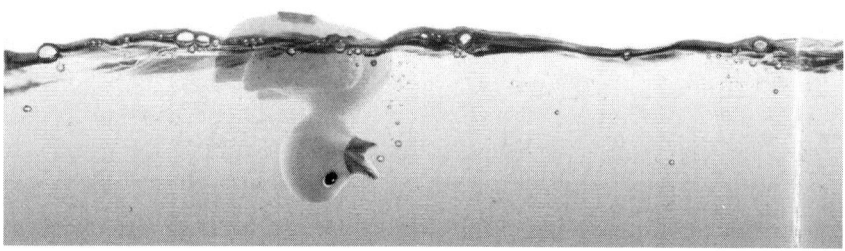

"Best time ever – it was an epic fail! I feel awesome!"

- Said no one ever.

Continuously we are told that failure is not an option. However, let's get real for a minute, you didn't pop out of the womb and start running a marathon. If you did, then you don't need to read this book. For most of us, we tried over and over again to walk or even crawl. While we were learning one of our first essential life skills, we fell a lot. We learned a lot from those early walking failures. Those failures taught us to crawl or walk without falling over. We were encouraged by our parent's hands grabbing ours and pulling us along, our family's St. Bernard's fur as a way to stand up, or a sturdy wall or two. Most of the time we had a soft landing on some thick cushy carpet which we should thank our ingenious parents for installing before teaching their child to crawl or walk. My rear end and knees thank you both Mom and Dad.

Business has traditionally told us from our very first day on the job that failure is unacceptable. You can't make a mistake, or you will get fired. There merely is no room for error of any kind. In business and sometimes life, avoiding failure is more important than succeeding. We may not have accomplished anything, but at least we didn't fail. We go to great lengths to avoid failure. We put ourselves into a make-believe world where no mistakes can be made or are even possible. We work ourselves to the point of exhaustion all in the name of not failing, yet we still fail.

Deluded in the belief that failure isn't an option, we are at a loss on how to handle failure when it does occur. Since there is no option to fail, there is not an option to fail safely. We ignore the fact that failure happens. It happens all the time all around us. Successful individuals and organizations realize that positively dealing with failure leads to better outcomes and more success. They have learned how to fail fast and fail safe. To understand why failing is so essential to your future success, we are going to take you on a journey to help you think, learn and work differently.

Together we will learn to harness the power of Failing Fast and Failing Safe with insights and models that give you the tools to leverage failure for both your personal and professional success.

Let's Take a Journey Together

Pictures are worth a thousand words. Let's look at three pictures that show the different levels of leveraging failure as individuals, teams, and organizations. Where can failure take you?

You Learning from Failure

Failures Occur

As individuals, we start with small failures. What we learn from small failures builds the foundation to handle larger failures.

Your Growth from Failure

As we grow from failure, we move to the next step of failure development by further harnessing failure and building new skills, characteristics, and self-esteem to strengthen our foundation to handle larger failures.

More Failure and Growth

As we progress forward even further in harnessing failure, we can enrich and fortify our foundation to handle failure more effectively. New leadership and self-awareness skills are built around a deep foundational understanding of how to harness failure for success.

Failures Occur

In a team environment where failure is avoided or not embraced, failure is often hidden or handled by individuals not collaboratively. The knowledge learned from failing is held with a single or a few individuals.

Growth from Failure

The team begins to start harnessing failure by supporting each other when a failure occurs, and lessons learned are more collaboratively shared.

More Failure and Growth

The team wields failure as a tool to innovate. The team produces successful results faster with greater agility and adaptability. The team is very collaborative on learning from failure and creating an environment where failure can happen safely and frequently.

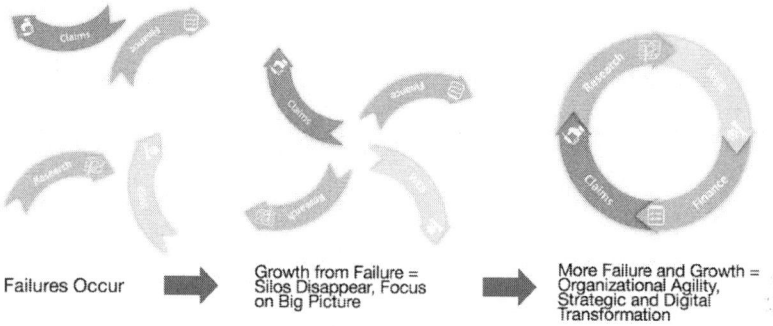

Failures Occur

In organizations, failure and fear are used to create silos of systems, processes, and knowledge. Failures are hidden or avoided entirely. Lessons learned from failure are not gathered or utilized to learn from in most cases. Organizational silos see themselves as political rivals and not as one cohesive organization. Each silo has separate goals and objectives that conflict when put into the perspective of organizational goals. Environments for safe experimentation and failure do not exist or are too tightly controlled.

Growth from Failure

Silos are stilling operating as individual units and begin to disappear as they are focused on a single set of goals, harnessing failure from effectively, and sharing lessons learned across silos. Failure is seen as an advantage to learning quickly. Lessons learned are sometimes shared between former silos and roles can move somewhat freely between former silos so that resources are more flexible and adaptable. Lessons learned are not always acted upon leaving important improvements left behind.

More Growth and Failure

The organization actively encourages experimentation and failure in safe environments. Lessons learned from experiments are shared through the organization with lessons learned. All levels of roles have an environment that allows experimentation and learning from failure. Roles are given time to pursue improvements based on lessons learned to build a stronger more robust platform for experimentation. The organization has a clear set of goals and all parts of the organization are active in achieving those goals.

Our journey to Fail Fast and Fail Safe will take us through many including:

- Understanding Failure
- Learning from Failure
- Curiosity and Failure
- Challenging Your Culture to Fail
- The Fail Fast Model
- The Fail Safe Model
- The Fear Model
- Agile and Failure
- Putting Fun into Failure
- Thinking, Learning and Working Differently to Fail

Before we dive into the topic of Fail Fast and Fail Safe, let's look at how failure is perceived in our personal lives and organizations.

What is Failure?

We start with the concept of failure, what failure means, how failure is seen as negative and presenting the importance of being adaptable. We must define failure and understand it on a different level. The Merriam-Webster dictionary defines failure as:

"Failure is the state or condition of not meeting a desirable or intended objective and is viewed as the opposite of success." [1]

When we think about failure, we see it as a negative thing. We avoid failure at all costs. Is failure a bad thing? Can't we learn from it? Let's change that definition. Our definition is:

"Failure is the state or condition of not meeting a desirable intended outcome or objective that creates an opportunity to learn and grow assisting in the attainment of success."

That new definition is powerful and starts the process of thinking differently. However, the general public views failure as individuals and organizations as:

- No common sense. Everybody knows that will not work!
- Stupidity. That's the dumbest thing I have ever seen or heard.
- Bad Decisions and Judgement. They weren't even thinking.
- You didn't think it through! I saw that one coming a mile away. Why didn't you?

---

[1] Merriam-Webster, merriam-webster.com, failure definition, September 11, 2018

It's clear these widely held beliefs of individuals and organizations drive us into the false impression that failure is bad. Hearing this criticism from others is demoralizing. Failure is something to be avoided rather than to be learned from.

"Failures are just weaknesses turning into strengths."

- Bob Prentiss

Stop Seeing Failure as Negative

When we fail we stop trying and give up, so we don't have to face our failure. We don't want to experience any of the negative things that failure causes us to feel. We avoid failure by not even attempting anything new or innovative – it just becomes too risky to attempt. When we don't fail - we don't learn. Giving up diminishes our ability to learn from failure, and we diminish our learning experiences.

Failure can also help us learn and move forward. You can't learn if you don't allow yourself the opportunity to learn from failure. By framing failure differently, it can be seen as an opportunity to learn and grow. Letting the fear overtake us, we lose sight of what we can accomplish.

Starbuck's Wikipedia article [2] and their website [3] tell a great story about turning failure into success.

---

[2] Wikipedia, en.wikipedia.org/w/index.php?title=Starbucks&oldid=858832399, Starbucks, September 9, 2018
[3] Starbucks, www.starbucks.com/about-us, August 11, 2018

Starbucks started as an Italian Coffee Shop in Seattle, Washington in 1971. The menu was entirely in Italian. Not one word of English. Italian Opera Music was played endlessly in the background. There were no chairs and most of the restaurant at the time was standing room only. Starbuck's didn't make a profit for the first nine years of its existence. Unless your goal was to cater to your 10 opera-coffee-loving friends, by all accounts, Starbucks was a failing business for a long time.

Starbucks didn't quit and decided to take a new path. It took Starbucks 9 years to be an overnight success with stores worldwide. Not being fearful of looking at failure and learning from it, many changes were made to build the brand.

- Menus are almost entirely in English at Starbuck's these days with a few Italian terms like "Venti" and "Grande" remaining on the menu. Latte, Mocha, Cappuccino all became everyday words due to coffee shops like Starbuck's becoming more and more mainstream.

- They added more seating so that you could enjoy your beverage more comfortably.

- Adding Wi-Fi also helped in making the atmosphere friendlier. It created an environment in which you could purchase a coffee, connect to the world via your laptop, or work a little away from the office.

- Opera music gave way to Coffee House music that arguably was made famous by Starbucks and other coffee houses. Did you know that Starbucks has its own record label? Yes, it is that popular.

- New drive-through windows made it easier for over-caffeinated drivers to grab their favorite beverage. Making lousy hair days a little more bearable.

- Starbuck's expanded their menu to include many new items like blended beverages. Even sushi and wine are on the menu in some of their stores.

This company now has over 10,000 stores worldwide. Starbucks didn't burst out of Seattle a rock star. Instead, it took years of failure and learning from it to craft their brand. As an investor, would you wait nine years to get a return on your investment? Probably not. Having the tenacity to keep failing and keeping learning from the failure builds success. You can't succeed if you never even attempt to fail.

Being flexible, adaptable and open to change is a crucial ingredient for success. Ideas and projects will need to change in their scope and context to apply the lessons learned from failure. Don't get caught up in not being flexible and holding to an idea or goal that can't work which can lead you to play the "Emperor's New Clothes" game. More about this failure game in a few pages.

Are you ready to see failure in the right light? Let's test that out.

Reaction Test

Look at the following words and then document your reaction. In the "Reaction" column you will score the word based on how you perceive the word. Do not fill out the third column yet (score) – leave it blank. Do not proceed until you have filled in the reaction column (spoilers ahead!). Grab a piece of paper right now if you don't want to write in your book. Hint: don't overthink it – this is not a certification test.

Reaction Value Table

Numerically value your reactions on a scale from 1 to 3.

Value	Your Initial Reaction
1	Positive
2	Negative
3	Neutral or Indifferent

Step 1 – Complete the Reaction Column

Read each word in the "Word" column. Using the table above, enter the value for your reaction. If your reaction to the word "Cancel" is positive, then enter "1" in the "Reaction" column. If your reaction to the word was negative, then enter a "2" in the column you're your reaction to the word was neutral or indifferent, then enter a "3". Don't put anything in the "Score" column – it will be filled out in the next step.

Word	Reaction Value	Score
Cancel		
Re-plan		
Start Over		
Nullify		
Annul		
Modify		
Substitute		
Rework		
Change		
Reorder		
Uncertain		
Ax		
Abandon		
Scrap		
Alter		
Tweak		
Replace		
Withdraw		
Countermand		
Retract		

Word	Reaction Value	Score
Invalidate		
Fail		
Void		
Redo		
	Total	

Step 2 – Score Your Reactions

Convert all of your reaction values from the "Reaction" column into scores. Using the table below, take the reaction value and determine the scope. Write your score into the "Score" column.

Reaction Conversion Score

Reaction	Score
1 – Positive	3
2 – Negative	1
3 – Neutral	2

If your reaction score was a "1", write "3" in the score column.

If your reaction score was a "2", write "1" in the score column.

If your reaction score was a "3", write "2" in the score column.

Example

Word	Reaction Value	Score
Cancel	3	2
Re-plan	2	1
Start Over	1	3

Step 3 – Total the Score Column

Add up all the values in the score column to get a total. Write the total in the total box at the bottom of the test.

Step 4 – Understanding Your Score

Your Score	Your Perspective of Failure
1 – 23	Negative Nancy or Ned
24 – 46	Moderate Molly or Max
47 – 72	Positive Paula or Phillip

All of the words listed in this exercise are typically seen as negative. People often feel uncomfortable with these words. If you are going to start failing fast and failing safe, we have to look at them as positive. Why? Negativity wears us down; it keeps us from moving forward. We quit. If you quit, you will stop learning. Learning is vital, and we must learn every day. The day you stop learning is the day you (or your organization) dies. The approach we should take with these words is a positive approach.

If we are told that we should "scrap" it, our response should be "Awesome! Let's chat about which pieces we should scrap and why, then scrap them, and find a better way to do it." If we are told that we are going to be "countermanded," our response is "Awesome! Let's discuss why and then find a new pathway to success." If we are told someone is "uncertain" about our approach, our response is "Awesome! We have an opportunity to discuss with them why we are doing what we are doing and if it is not the best option, we now have a better context for moving forward."

Did you notice the theme in all of our responses? They all include options to discuss the context for the decisions. Blindly following decisions does not lead to healthy failing fast and failing safe. Lemmings running for the cliff is not a great practice. However, rather than complain, bemoan, reject or avoid these negative words, we can embrace them as positive opportunities for failing fast and failing safe.

Failure and Abandonment

Starbuck's had tenacity. However, even with tenacity, you may need to abandon an idea because it doesn't make sense, or it is not feasible. The abandonment occurs when after several attempts the lessons learned from failure point directly to the uncomfortable fact the concept isn't going to work. Failure still needed to happen to understand that you needed to leave the idea altogether or change the idea. Being open to new paths to create the idea is critical. Flexibility and adaptability are the key drivers in being successful.

Leaving or abandoning something you love isn't easy. Making this decision when the concept is well-liked is extremely difficult. It's important to be open to the fact that an idea is just not feasible. It is certainly true when it no longer makes sense. The feelings that accompany failure are powerful in this situation.

Lessons are still learned even if you abandon an idea. Lessons learned are a great thing to document, but if you don't act on those lessons learned the meaning is missing. You can't learn from the failure until you reflect upon that failure in an unbiased way. Not acting on what you learned from failing dooms you to repeat those same mistakes with future ideas and projects. Reflecting on your failure is where you begin to conceptualize why the failure occurred and how to learn from that failure.

Failure and abandonment discussions must occur and should not be avoided. Discussions of this nature are uncomfortable. Prepare your conversation by gathering the data, factual observations, and lessons learned on why the idea failed. Think skeptically about comments and observations from you and others. Don't take them at face value and try to understand differing points of view. Try to keep emotion out of the discussion which can be difficult when dealing with failure. Remember those negative feelings you have with failure? The entire group of people you talk to will be having those same feelings. Acknowledge everyone and their point of view.

Focus on the data and the lessons you learned to help address the feelings of the group. Ask for feedback and validation of the data, facts, and observations. Ask for any directional changes the group thinks might work to put the concept on a more prosperous path. Thoughtfully review previously attempted changes to make the concept work. Understand why those changes failed or not. It's not easy for a group to quit an idea. Keep them focused on the verified data and observations used to create the lessons learned. Be prepared to decide to abandon the concept and possibly create a different one. Maintain flexibility as much as possible. Be transparent with your data and don't hide any relevant data or facts because they would make the conversation more difficult. When it's discovered you hid those facts, it will break your other's trust in you.

Times When You Can't Fail

There are times when failure isn't an option or necessarily ideal. An elevator that zooms up to the 50th floor of an office building is a good example. Landing a commercial aircraft and open-heart surgery are other great examples of situations where you can't fail. Both of your authors are not fond of air travel so let us be clear, there are times when failure is bad. The irony? One of us travels about 70% of the year and is very passionate about airlines getting it right. It's essential to recognize critical situations where failure can't happen. Nothing is perfect, and everything will fail at some point even something as critical as open heart surgery. You need to ensure you have a soft place to gracefully land and recover in the event failure happens. For elevators, they have a mechanism that ensures an elevator doesn't uncontrollably plummet to the ground. If you have been trapped in an elevator between floors (as we have), you know first-hand there are safety precautions in place to prevent the elevator screaming down the elevator shaft to the ground floor. These safety measures allow the elevator to fail safely. Okay, modern buildings do so, in old buildings take the stairs!

Before leaping out of that airplane have a parachute. Then a backup parachute. Test the primary and backup parachutes on the ground to ensure they are in working order. Better yet test both primary and backup parachutes on the ground and have a back plan for another person to save you if those fail. Make sure the person saving you also has a primary and secondary parachute. Are there airbags for that sort of thing? Having 2-3 backup plans to land softly is worth the effort (sorry – just air travel paranoia there). A little planning ahead is a good thing. There is such a thing as over planning, and that will take us away from learning from our failures. Most of us don't have to worry about parachute not opening and crashing to earth, but we do have essential tasks we perform every day. Having a backup plan when things so wrong can save you a lot of stress especially for the things that cannot fail.

What is Failing Faster?

Failing faster is about creating situations and environments in which an individual or organization can quickly prototype, demonstrate, test, and learn. It is a strategy of building a product by quickly getting feedback from users or customers, and rapidly adapting the design. Lessons learned are incorporated promptly into the solution design so that the process can start again quickly. Overly used in Silicon Valley startups, Fail Fast as a lot of buzz around it.

When there is a high level of uncertainty, it is often less expensive to start building a working product to learn whether or not the product is viable or feasible. Additionally, this method allows you to quickly get feedback from potential customers on the process or product. A benefit of this method and approach is to kill products fast before too much money is invested in their creation. Organizations don't want to spend significant time and money into a process or product that will not work or will not sell. The key to being successful with this method is small incremental steps.

Silicon Valley and startups use the phrase "fail fast, often fail" when talking about product development. These companies are looking to get something out there to see if it sticks as fast as they can. If it doesn't work, they can "pivot" or adapt to what they learned from their failure to try a new approach.

"Ready, Fire, Aim" is another phrase that is commonly used and associated with the failing faster. This method gets ready by building out a prototype (or "ready"), then puts the prototype out for the customer to test (or "fire") and learn about why it missed the target (or "aim").

Agile Methodology also adopts this method in their fail fast approach. Get something out there for the customer to interact with, get their feedback and then adapt. Other terms that have a similar meaning are "Fail Better" and "Fail Forward."

This approach can be used in our daily lives when we as individuals try out something new like for example trying on a pair of new sunglasses. We need to get new sunglasses because the old ones broke, so we head out to the store. You look through the rack of sunglasses to see which ones might work. You try them on and look at yourself in the mirror. You ask your friends or spouse for their feedback. Based on what you see in the mirror and the input you get, you then create an image of the sunglasses that will fit your needs. You discard sunglasses that don't fit your vision, and a new pair is tried on for size. You keep trying on sunglasses until you find a pair that looks the best.

This approach allows you to try a variety of sunglasses quickly. It also allows you not to choose any sunglasses and walk away from the store. You are abandoning the sunglasses effort because all the attempts to succeed in finding sunglasses that meet your needs failed. Remember those feelings of failure? To a smaller extent, you are feeling them in this case. You can always go to another store and the process of trying on sunglasses all over again. You might revise your criteria for the sunglasses or address if you need them at all.

What is Failing Safely?

Understanding how to fail safely is as important to understanding the importance of why failure helps us learn.

A few years ago, on a server migration project, the Database Administrator on our team announced in an impatient tone, "This database can't go down! It's mission critical." That was an interesting little bomb to drop on the conference room table, so I asked, "Why can't it go down? What makes this database mission critical to the organization?"

The Database Administrator elaborated further on the database purpose and its critical nature to the organization. There was a backup done every night, but no one had ever tested the restore process. There really should be a commandment in technology that says, "Thou shalt not backup or archive data without ensuring the data can be retrieved later."

The team decided that it was too risky to move a database from one server to another without first ensuring the data would not be lost. The group performed a full backup and then attempted to restore it on the new server. A significant disaster occurred as the new server configuration did not match the old server configuration. We were still okay because the old server was still in production and there was no disruption to the business teams accessing the data. The team wouldn't switch over to the new database until the restore process worked.

After a dozen attempts and countless hours of documenting the steps needed to move from one server to another, the team was successful. A dozen attempts ended in complete failure. Even though we failed repeatedly, we had a safe place to land.

After each failure, the team got more and more scrutiny from management. Management had a tough time with how long the migration was taking. Management expectations were that the migration would take a few days. They conducted more and more meetings. Management kept asking, "Why didn't you see that coming?" We replied, "We have never done this type of migration before." This led to putting all new players on the team that was more experienced so that the migration could move more quickly. Did that migration go faster? Not at all. The new players didn't experience the failures of migrations we previously attempted. The new team proceeded to make the same mistakes the old team made. The new team's strategy was to keep pounding on it until it worked.

That didn't quite work out so well. After 36 hours of non-stop troubleshooting, the new team was no closer to solving the issues than the old team. It took them two days to recover from pulling a marathon 36-hour shift. We had another meeting with management, and they too reported they made no significant progress.

We sat down and set management's expectations more clearly. No silver bullets or SWAT teams are going to speed up this migration. The teams together created a list of lessons learned. The document outlined what went wrong on each attempt, the root cause of why it occurred, and the steps we took to make sure it wouldn't happen again. After every attempt, the team got together and went through our list. In the beginning, it was excruciating, but as we progressed, we grew more dependent on the list as a tool to help us move past errors and even avoid potential future failures. There were a lot of "chained" events that were discovered using this technique. You must do this step first and then this step second. Logical GO and NO-GO decisions got added to the implementation plan.

In the end, we had successfully moved the database from one server to another. Our list became a standard that other database moves would later follow.

Getting Past the Word Failure

The word failure has a lot of bad connotations, and many people don't like to use the word even if it means we learn and therefore is something positive. Say the word failure, and you will get an adverse reaction every time. To get past these impressions, you need to frame the word as something that is just natural and part of the process. Find another word that makes you comfortable as long as you understand the meaning behind it is still failing or missing the target.

Creating the right environment for failure can be culturally tricky. In some cultures, you can't even say the word failure without being insulting. Nobody will let you say the word failure in a meeting. In this situation, you might need to change the word to avoid adverse reactions. Find a word that will work for the culture. The label for failure may vary, but it still has the same meaning as failure.

Just replacing the word failure with success creates an impression that we need to have a high level of perfection before it can be released. This perfection defeats the ability to learn. It also slows down the speed at which you can learn. You are asking for feedback and valuable insight by putting out a working product with which that your users or customers can interact. By focusing too strictly on success outright with complete perfection, you won't be able to get the valuable insight from your users and customers. Ideas come from many different perspectives and those perspectives strengthen your product.

Find a word that will work for your organization but understand the concept of failure. You may label it differently to get a more positive reaction, but it has the same meaning nonetheless.

In today's business world, nobody throws a product out there without some testing even if the product is tested entirely within your organization. The feedback from testing that is performed by only internal staff limits the amount of perspective or feedback you can get on the product. The broader the audience, the more likely you are to get broader perspectives and feedback. Stronger feedback gives you stronger ideas for creating an excellent product.

"We didn't fail – we found defects in the application code!" I would argue those defects found in testing are in failures and that your testers will learn from those failures. They will even adapt to work around those failures. A meets the definition of failure:

"Failure is the state or condition of not meeting a desirable or intended objective."

If it doesn't work in the technology world, it is labeled a defect. A defect causes the product not to meet customer expectations. The business side of the world doesn't typically use the word "defect," but it is applicable for use on the business side. Customers interacting with a product at an early stage quickly uncover defects. These defects can, in turn, generate new features and capabilities previously not discovered.

Embrace Failure and Stop Fear

Many start-ups and growing companies talk about embracing failure. They repeat the Silicon Valley phrase "Fail fast, fail often" like a mantra over and over. It doesn't necessarily mean they are not afraid of failure or that they will accept it. Speed to market for a fledgling start-up company is mission critical. Start-up companies are always in a battle to show their venture capitalists they can quickly turn a profit.

Fear is a strong motivator for any organization. The practical way to combat the fear of failure is failing in small increments, not large ones. Blowing through a million dollars on a prototype only to discover it wouldn't work for the organization isn't acceptable. Investors will set funding constraints for prototypes out of the desire not to have their funding exhausted on products that are not unfeasible. The product may sound great on paper, but building it is an entirely different matter. The practical approach is to quickly create a prototype to demonstrate the product can be realized and profitable.

There are no shortcuts here. Doing the work is still required. No magical formula can ensure a prototype is feasible. Take shortcuts where you can to deliver quickly but be aware those shortcuts can cause you to overlook possible opportunities to learn about undiscovered capabilities.

The term "hacking" implies creativity, imagination, and a relentless drive to solve a puzzle. It also suggests cybercrime and guys stealing your credit card information. If you have a genuine commitment to embracing failure and learning from it, you won't feel a need to take shortcuts. Taking shortcuts that do not allow you time to think or research get a negative response. You will recognize alternatives as either bad in that they introduce noise into solving the puzzle, or useful as they are discoveries of doing the work differently and faster.

Failing Fast Aftermath

Figuring out what works means you have to shut down a lot of ideas and there are many products that get significant criticism from your users and clients. It can be demoralizing. The best way to soften the blow of this criticism is to have everyone in the right state of mind.

Set expectations that new ideas are valued but encourage those ideas to flow. Ideas can come from anyone in your organization, the team, or customers of the product. Just because an idea is shot down doesn't mean it didn't spark other ideas or won't be addressed in the future. Keep a list of all the product's ideas and the outcome of those ideas. Think of it as a capabilities list that is tracking all the potential capabilities and whether or not added to the product. With many ideas swirling around it's hard to keep track of them all but keeping them together in a list has benefits. The list ensures ideas are not lost or don't get brought up and shot down all over again.

Encourage the organization and team to keep giving ideas even when it seems there aren't any more ideas. Keep looking at things from different angles both internal and external to get a broader view. By creating products with small enhancements over and over, you give more opportunities for customers to provide feedback on your product. Daily product updates are confusing because your customer can't get their head around what it is you are trying to create. Choose a product update or release schedule that make sense for your product. Updates to a software product usually find their way to customers once or twice a month. Heavy equipment or durable goods might only update their product once or twice a year.

Positively acknowledge ideas. When an individual gives an idea, they are putting themselves out there. Criticism or negative responses to ideas will shut down new ideas from being created from the entire team. If the team observes consistent adverse reactions to new ideas, they will no longer offer new ideas.

Recognize that after a stream of failure your team that is building the product can lose momentum. Don't always focus on the negative. Look for the positive feedback. Show the team the journey has been incredible with everything that we have built and learned about the product.

Customers can be critical for the sake of being critical. Ignore the overly critical and focus on criticism that is constructive to keep your team from being beaten down by critical feedback. Read all the feedback but understand that the purpose of feedback is to understand where the product can be improved rather than torn apart.

Build Small, Be Smart, Pause and Learn, Build Again

Fail fast isn't about the significant failures or changes, it's about little ones. It's an approach to running a company or developing a product that embraces lots of small experiments with the idea that some ideas will work and grow while other ideas will fail and whither.

- **Build Small**. Make small incremental improvements with each version of the product. Avoid trying to boil the ocean in one large increment. Focus on a specific function, feature or capability of the product for each version.

- **Be Smart**. Verify incremental improvements are not overwhelming for the team to analyze and build. Make sure they are realistic to achieve. Break up large functions, features or capabilities into the smallest pieces as possible. Look for dependencies been small steps, so they will logically progress forward step after step. Verify how everything can be chained together to understand the best way capabilities can be added incrementally to the product.

- **Pause and Learn**. Allocate enough time to test the product with customers for more comprehensive feedback. Take the time to reflect and learn. Look for better approaches to delivering the product to the customer.

- **Learn and Repeat**. Take what you learned from the feedback and build it into the product. Track what you learned and new ideas.

"I've missed more than 9000 shots in my career. I've lost almost 300 games. 26 times, I've been trusted to take the game winning shot and missed. I've failed over and over again in my life. And that is why I succeed."[4]

- Michael Jordon

The Failure Games We Play

There are many games that individuals and organizations play when dealing with failure. Some are somewhat complicated, and others are easy to play. Each of these games shows how individuals and organizations deal with and learn from failure. The games listed below are the games we shouldn't be playing. Later in the book, we will be talking about collaborative games we SHOULD be playing.

The Blame Game

When failure happens, our first instinct is to find someone to blame. We want to save our reputation and save face. Failure starts the political game of Blame Storming, and human beings have raised the bar to new heights on how this game is played. Instead of the direct

---

[4] BrainyQuote, www.brainyquote.com/quotes/michael_jordan_127660, Micheal Jordan Quotes, August 1, 2011

approach of, "This is all your fault! You suck!", we use the subtle approach of, "I have concerns about your credibility" or the ever-popular "This certainly is a hit to your credibility and career here at this company."

Magnificent dodges and attempts to blame other ensue leading to, "I'm not taking the blame for this," and the blamestorming game starts all over again. The game finally ends with the "It's a very complex situation" play or no other victims could be found to blame.

Being clever it was discovered how to play the blame game at a whole new level. The realization was that you only needed to bring up a name and blame them. No one checked to figure out if that person was real. This factious person got the blame for many things. The team went even or far as to create an email account, populate a cubicle for this artificial team member, and display fake family photos on the desk.

In this social experiment (never repeat this experiment), we blamed our poor helpless yet fictional team member for many failures. Managers and directors would avoid understanding the failure and demand that some action is taken on this team members poor performance.

The division leader at the time even went so far as to write up a performance plan or "shape up or we will fire you" document for our fictional team member. The divisional leader demanded a face to face meeting with our fictional team member. Let's say the meeting with the divisional leader informing her of the fact the team member didn't exist did not go well. In the many meetings that followed, we had some terse and uncomfortable conversations on the organization's desire to find someone blame. Then fire them rather than considering the issue and determining a root-cause.

The strategy for winning this game is not to attack an individual but rather the systems, processes, and data that are causing the failure. Focus the group on the how or why of the problem at hand – not the who. Don't allow others to pile up on an individual or team.

When you hear, "I think John's team is responsible for this mess." Immediately ask, "What systems or processes are involved here? Are we just jumping to blaming without understanding the failure? There are several teams involved here, and we need to focus on getting to how the failure occurred or why the failure happened. It's not about which team to blame – it's about understanding the failure itself." It's important to address the need to blame an individual or team quickly. Given any chance, we will let this blame game rapidly grow out of control. Nip it fast when you hear this game starting up.

The Emperor's New Clothes Game

The game is called "The Emperor's New Clothes." The game is quite simple to play. A major directive is established by the board of directors or a manager that isn't feasible or based in reality. The initiative starts and encounters failure after failure. No one ever says the initiative has problems publicly. All communication about the initiative demonstrates how wonderful things are going without any major issues. The initiative is released, and no customer ever buys it. It's an epic failure. Everyone points and laughs at it privately. No one will tell the emperor he failed. Typically, the Blame Game starts after the initiative fails. Rinse and repeat. Start another initiative, play the

"Emperor's New Clothes" game, fail and fall flat, then play the "Blame Game." It's a perfect cycle that it is hard to break.

Not embracing the fact that we are failing or have failed is a recipe for great disaster. An organization or individual that fails to understand their failure is doomed to repeat it. Failure is an opportunity to learn and improve. Failure is an opportunity to strengthen the organization's ability to be more flexible.

The way to stop this game is to thoughtfully put together the facts and design for the proposed initiative before it enters any construction phase. Create a pro and con list that is an honest and transparent assessment of the initiative. Carefully prepare a meeting with decision makers and leaders. The objective of this meeting is to ensure their vision for the initiative is fully understood. We are checking in to make sure we are on the same page with your leadership. Present in an unbiased way your findings and ask for their thoughts on the initiative. Are there topics or knowledge that is missing? How can we ensure this initiative will be successful? Ask the open-ended questions to allow your leadership time to think through the initiative in entirely.

The Two-Sided Sabotage Game

Next, comes the "Two-Sided Sabotage" Game. This game is quite similar to the "Emperor's New Clothes" game, but with a twist. The intention is not avoidance but rather sabotage. A new initiative is announced, and the leader of that initiative goes around asking for support from his directors and managers. Everyone knows the initiative was strong political clout behind it. So, arguing about whether or not it should be done is considered impossible. Our initiative leader will hear a director say, "I support you 200%", but what they do is either put the most undesirable team members on the initiative or under staff the initiative. One side is always happy and supportive while the other is intent on sabotage while looking supportive.

In this game failure is the weapon of choice. Using failure as a wedge to build discontent among team members instead of embracing and learning from failure create a perfect environment to play this game.

Our saboteur needs to start placing doubt in everyone's ears and not directly bringing issues to the initiative leader. Convincing other teams to avoid communicating issues creates the illusion of no one cares and thus builds the groundswell of discontent. In this discontent, the saboteur reigns supreme. Once the groundswell is massive enough, it's hard for the initiative leader to push the initiative forward. Every step is a painful one. After a short time, there are just too many issues to address, and the leader of the initiative spends more time problem solving than building the initiative. The initiative most likely goes back to the drawing board or is canceled. The saboteur wins. If the initiative leader pushes it forward, you are now playing the "Emperor's New Clothes" game.

To win this game, you must take away failure as the weapon of choice. Publicly embrace failure and issues by communicating them widely and openly. The saboteur has the upper hand when there is discontent. Create the environment in which team members see the issues and failure points they bring up. Make sure those concerns viewed as opportunities to improve the overall initiative. Know your stakeholders, managers, and directors. Interview them to privately and show them that failure isn't the problem. The problem is not learning and changing course from the failure.

Ensuring victory in this scenario will mean building up champions for the initiative. You will need quite a few champions to ensure they can be heard above the noise the saboteur creates. The object is to create more champions and supporters for the initiative. You really can't have too many. To be effective, you will need to engage potential champions and advocates one on one. You'll need to determine if they really are champions and supports or just giving lip service.

Spending significant amounts of time on someone who is firmly entrenched in stopping the initiative is unproductive. You will not be successful in attacking a saboteur's position publicly. Most will view your position as too aggressive and mean-spirited. Be careful to listen to the saboteur's concerns and understand them. Avoid being defensive – seek to understand. You should not assume that all points of view have been presented. The saboteur may have some piece of knowledge that is hidden, so it's best to try to elicit their thoughts on the initiative in a non-threatening way. The objective is to discover these hidden pieces of knowledge that are significant impacts to the initiative. That will require building their trust in you to disclose this information. Position yourself as understanding.

The Marble Pedestal Game

The "Marble Pedestal" game is rare game to play. The game revolves around the fact an individual or organization believes it can't fail – ever. They will say that the organization is too energetic, entrepreneurial or smart to fail. The fundamental belief here is the idea that they are flawless and at the height of perfection. In this game, the belief drives the individual or organization to put its blinders on and avoid even remotely learning from their failure.

In an organization or team, the game is played by everyone from management all the way down the chain of command. "We are too big to fail." "We have been a market leader forever, and that is not going to change." "How can we fail? We are the best company ever!" The organization doesn't' believe it needs to learn from failure because failure does not happen. With every win, the organization endorses the flawless belief and continues to raise the organization on the marble pedestal higher and higher. There is never any reflection back on the win to see if there were failures that had lessons to learn. They never look back because they are on the marble pedestal looking above everything.

Failure for organizations playing the marble pedestal game is disastrous. The organization was so high up on the marble pedestal that the fall to fail hits them hard. They were never prepared for failure. The organization falls apart, productivity crashes and morale bottoms out. Avoiding the big crash is easy by creating a soft place to land. If even you have never failed, planning to land someone safe when you do fail isn't something that comes to mind. Make sure you have a safety net. Embrace the failure and learn from it. An organization's unwillingness to not embrace failure takes away the organization's opportunity to learn from success or failure.

The marble pedestal game is a cultural game that is difficult to win. To win this game you will need to become a champion of failure and how failure helps organizations learn and adapt more quickly. Cultural change takes a significant level of energy to pursue and will need support from your leadership. Position yourself as someone who desires to utilize the principals of agile. Talk about your desire to embrace Agile. Also, talk about your willingness to help your team be better agile players. Show your team the advantages of failure and show strong leadership when someone fails. Responding to failure in a supportive way will show your team that you are open to taking reasonable risks and failing safely.

The Risk Mitigation Game

Failure and risk have a lot in common a project. Both are roadblocks to the project's success. Both should have a mitigation plan or a soft place to land planned out in case they occur. Failures occurred in the past, are happening immediately or will happen in the future. Risks tend to be future focused or potentially could occur. In some cases, risks that occurred in the past are brought into the current risk plan with mitigation plans.

In business and project management it is a best practice to plan for risk. Organizations recognize that failure might occur, and they need to prepare for it in their projects. Project managers perform risk qualification and quantification on their projects. In most cases, the risk management plan is devised at the start of the project shortly after the funding approval to start the project. Typically, most project managers focus entirely on those risks that are within the boundaries of their project and rarely look at organizational, product, market, or other external type risks.

The purpose of the risk plan is to find that soft place to land by identification of potential risks, determining which ones are the most likely to happen and which ones have the most significant impact on the overall project success.

The game is played on projects frequently. At the start of the project, risk identification is performed. It is here we create the illusion that our risk plan has identified all potential risks. All potential risk includes those that are unknown or that we didn't think to identify. Looking at the risk plan, do you see a risk called unknown? Every project manager understands some risks can't be foreseen. When an unknown risk occurs, there is no plan in place on how to address it or act on it to mitigate or resolve the risk. By not including Unknown as a risk on the risk plan, it creates the illusion that all risks have been identified and managed.

The game is further played by create the risk plan once and then put it on a shelf never to be updated or seen again. Frequently senior leadership limits the risks on a status report to the top 5 or top 10 risks. The focus is entirely on these small sets of risks and not on the larger picture. Focusing on a smaller set of risks is done because it's too complicated to go through every threat identified in the risk plan. Surprise risks are quite common on projects. Some of these surprises were accounted for in the risk plan, but they were far down the list and never got any focus. Unforeseen risks lead to fire-fighting tactics by the project manager to develop a mitigation plan on the fly. By not continually looking to see we here you are failing, you are not looking for a soft place to land. By not looking ahead everyone is in react mode and is not proactive in actively looking for potential risks or failures to occur.

We have some risks that appear on all projects but do nothing about it. We don't create a soft place to land when these risks or failures occur but instead accept that they will happen. "I will burn that bridge when I get to it" mentality falls into place. Reactive thinking and not looking ahead. Being practical and understanding that looking forward ahead of the team is essential for any leader.

A variation of "We don't' have any control over that, so if it happens we will have to deal with it" mentality. Declaring a lack of ability to control a risk prevents any planning on what to do in the event the risk occurs. Even if a risk is accepted, the mitigation plan needs to find a soft place to land in the event the risk happens.

Another dimension to this game is not verifying the mitigation plan. No one ever tested the mitigation plan will work. The scenario is that we find a mitigation plan and call it good. No one ever actually tests the mitigation plan to see if it works. When the risk or failure happens, the mitigation plan is executed and falls face first driving the team into reactive mode. Finding a safe place to land is vital before you encounter the risk. Testing that soft landing can take time and effort, but verification of a good place to land in the event of failure is a good thing.

Discuss and put in place a checklist on projects and sprints that can help them fail safely. This can take the form of setting up environments, configuration checklists, checklists for building and construction and sharing implementation plans that include tasks to help ensure things fail safely. Something as simple as a back out plan in case an implementation fails can be beneficial.

The King's Court

Steve Jobs once said, "A company is more successful if it values ideas over politics." In the King's court game politics are rampant and overwhelming the organization. Everyone is trying to get the king's favor so not to be laid off, reassigned or fired outright.

In the King's court, it's all about looking good in front of the king. Saying or doing anything to ensure your position is not in jeopardy lest you lose the favor of the king. In this game, any new idea is never able to grow unless it is a weapon of choice to defeat your rival or secure greater favor with the king.

This game has been played for centuries. In today's world, it's the CEO or another top leader in your organization that everyone seeks favor from. Everyone wants to look brilliant and wonderful. Failure under any circumstances is used to diminish other's positions and hopefully raise your status by being the one to point out the failure.

This game is subtle and played in almost every organization to some extent. Favor may not be desired from a CEO, but it could be a manager to which you directly report.

Making this cultural change from politics to ideas is exceptionally difficult. Such a change needs to occur on the top of the organizational structure and work its way down the chain of command. Organizations like these become incredibly bad at innovation and new ideas. It becomes too easy to create rivals that want to see you fail because you are innovative.

Winning this game requires you to have many champions that are fully committed to the idea and innovation. Creating those relationships in these types of organizations is very time consuming and slow. At every turn, there will be colleagues that will be trying to make you look unfavorable by picking on every little detail they feel is wrong. Failure of any kind is a blessing to tear your ideas down. You must build strong and broad coalitions of trusted colleagues across all areas of the organization. Be careful not to offend or avoiding involvement of key players that hold court over their areas of expertise, systems, process, or resources. Thoughtfully work with all our supporters, so it appears that they are all leaders of the idea.

Be prepared to only move the dial towards acceptance of the idea a little at a time. It may take months or even years before an agreement is reached. Be prepared that the idea may stall entirely, and you will not be able to move it forward any further. To win takes time and persistence.

It's Time to Play a Different Game

Let's stop playing games with failure. We need to understand that failure is an opportunity to learn and grow. We should put ourselves out there fully with the understanding that gives us the opportunity to learn, grow and be stronger. Try to challenge and change the games that are played above. You may even discover lots more games that are being played. Be on the lookout for them.

The game that we want to play is the learning game. Being in a position or place where you are always challenged to learn new skills and systems is an incredible experience. It builds a passion for learning and the understanding that learning requires failure. You can walk in knowing you are making a difference. The energy builds on top of itself. Your days are more rewarding and exciting. Work becomes fun.

Ideas are built from curiosity and exploring. Once new ideas are generating in volume, there are more options on the table of innovation. This gives an organization a lot to choose from when looking for that next innovative idea. No organization can stand still and not develop new ways to perform better or deliver new products. They become stale and wither away. Ideas and innovation are the lifeblood of any company.

Learn more faster and innovate by harnessing the power of failure. Encourage curiosity in yourself and others. Explore new ideas every day on your own or with your team. Change the mindset that failure is necessary to learn.

Chapter 2

Learning from Failure

A reporter was interviewing a President of a Bank on his success. "Sir, what is the secrete of your success?", the reporter asked. "Two words.", the banker replied. "And, Sir, what are they?", asked the reporter. "Right decisions.", stated the banker flatly.

The reporter asked, "And how do you make right decisions?" The banker stared directly at the reporter and said, "One word." "And, Sir, what is that?", the reporter questioned. "Experience.", said the banker.

Still holding the bankers stare the reporter asked, "And how do you get experience?" "Two words.", said the banker as he leaned back in his chair. The reporter leaned forward and asked, "And what are they?"

"Wrong decisions," stated the banker with a grin.

That sums it up pretty well don't you think? Wrong decisions are equal to experience which means we learn from failure.

Do you remember what it was like when you were a child learning for the first time? Has that experience stayed with you? Throughout this chapter we will discuss framing yourself as a beginner to learn faster and how to use different learning models to learn more quickly. We will also consider how an ancient Greek philosopher reasoned about what we can control and giving permission to ourselves and our organizations to fail. We wrap up this chapter by talking about creating habits for failing fast.

One key component of harnessing failure is the ability to learn from that failure constructively. Learning and failure go hand in hand. Not learning from failure produces zero gains as repeating the failure would be more likely in the future. You wouldn't know how to avoid it in the future unless you understood how the failure happened.

Let's use the example of learning to walk. When you fell and failed, there was someone there to help you get up on your feet again, and you adjusted your stance. These adjustments caused you to learn how to walk, and then walk with poise as you consistently adjust your stance. In your mind you were taking the feedback and the experience of failing, reflecting on that failure, and figuring out better approaches and then engaging those approaches in learning how to walk. Were you doing that as a small child? Yes, yes you were. In some cases, the feedback was extreme because we got hurt creating a more visceral response to the failure. Like the time Bob ran into a solid metal pole. No comment.

If you didn't adjust, you would continue to fall. You need flexibility and adaptability to improve. Trying out different approaches and seeing the results helped you figure which adjustments worked the best. Change is another significant component of failure. Strict adherence to not adjusting your stance will result in continuing to fail without any progress on learning to walk. If you always do what you have always done, you will get what you have always got. Being adaptable and flexible leads us to the next component of learning from failure – reflection.

Individual Reflection

Reflection on failure allows you the ability to look critically and skeptically at the failure. It helps you determine the cause of the failure. Looking for the cause of the failure will enable you to create potential ways to adjust. In the technology world, we sometimes refer to sprint reviews, retrospectives, post-mortems or lessons learned events. These are used to reflect upon the processes and issues we encountered to learn from them. As a group, these approaches are more formalized, but individual formality is not needed. Individuals can reflect on the failure by thoughtfully reviewing an issue or failure by themselves.

Reflection on the failure should lead to discovering issues that lead to the failure and actively addressing those issues that were discovered. All too often, however, we write up a list of things that we could have done better, and we don't do anything further. We don't dive into the cause of why they happened or how we can avoid them in the future. We don't get far enough into the details of the cause to help formulate a solid change.

Reflection isn't entirely about being negative or attacking. Reflecting on what went well is equally important. We want to repeat the good and avoid the bad as part of our reflection. Only focusing on negative things leads you or a group to a negative mindset. Add positive outcomes into the mindset. Think of reflection as a list of pros and cons. One side of the whiteboard is the bad stuff, and the other side is the good stuff. It is highly recommended you end on the positive. Positively finishing leaves you or the group leaving the reflection on a positive note. P.O.S.I.T.I.V.E.! We live our lives ¾ glass full. We always find a way to end on a positive note.

Reflecting Organizationally

Reflection in organizations is determining lessons that were learned. In Agile a retrospective is used. In other cases, a "postmortem" or post-project review is completed to compile of list of lessons learned. Unfortunately, organizations approach these opportunities as something to check off on a task list for a project. Organizations fail to take the lessons learned and apply them back to future projects which causes the same issues to appear again. Dismissing or not acting on lessons learned leaves those individuals involved in the activity feeling like working upon their experiences was not important enough for the organization. In turn, they don't provide much feedback in the future. Not every lesson learned is actionable, but taking the time to acknowledge and work on a few top priority lessons learned will help build a team's confidence and value.

Reflection allows the individuals to walk away with lessons learned that could be applied to their next project. Put aside time to reflect on your experience. Even if the experience was excellent and without problems, reflecting on it will still uncover new lessons learned.

Reflection leads us to change. Change is never easy as an individual or organization. Managing change is key to taking the reflections on the failure and bringing them into being. Not leaving them on a list that doesn't get considered for adjustments. The act of adjusting and changing is vital to learning from failure. If you or a group go through the process of creating a list of all the issues and failures encountered, but take no action on the list then what is the point? It just seems to be a waste of time. Nothing can be learned until you take action or make adjustments. So how do you do that? Here is a simple technique that will help you NiP iT in the bud.

NiP iT Technique

Do you have "those" meetings where a lot is said, but nothing is done about it? Well, it is time to nip it in the bud! It is not helpful as a facilitator of a meeting to let negativity rule or to let your attendees walk out with a sense of dread because they feel nothing will be done about the issues and complaints. This simple technique will help turn those frowns upside down. Nip It comes from the 'N', 'P', and 'T'. It stands for:

N = Negative

P = Positive

T = Task (To Do)

This is an approach to let people get their negativity out, then turn the negativity into a positive solution and the task needed to move it forward.

When should I use this technique?

Any time there is excessive complaints or a lack of accountability. This can be done in the middle of a meeting or if you have a planned approach for a meeting or workshop.

Preparation

- Whiteboard or flip chart
- Markers or pens
- Sticky notes
- Identify your time box for the effort. Shorter for fewer issues and longer for many. This could be a very long session if part of a workshop for a multi-million-dollar project.

How to Get It Done

1. First, we deal with the 'N' – the negative. Start with the conversation about what is not working, what is not helpful, what are the roadblocks, issues or problems. Do not try to solve it during the discussion. Just record each item on a sticky note or a section of your whiteboard. Example: Jon says "We don't have enough time to do the analysis. Why bother?" Have Jon write the following on a sticky note: "Not enough time to do the analysis."

2. Next, we turn the 'N' into the "P" – the positive. For every single issue that gets written down, you must find the positive – the opposite, the solution. For Jon's issue of "Not enough time to do the analysis." Together the team and Jon must now present solutions. There could be many solutions to a single problem. A prioritization technique such as 20/20, MoSCoW, Dot Voting, or other prioritization techniques assist in determining which solutions should move forward.

Example: One solution for Jon's issue is to hire more Business Analysts. It is typical to hear "Hiring more Business Analysts will never happen!" Well, it won't happen if you don't value the solution. More importantly, it won't happen if there is a lack of accountability to implement the solution. We ignore the comment of negative Nancy or Ned (sorry to all those really friendly, happy, and not negative Nancy/Ned's out there) and we must now make this go from the 'N' (Negative) to the 'P' (Positive) to the 'T' (Task).

3. What are the tasks we need to do to accomplish this potential solution? More importantly, who is going to own them when we leave the room? Example:

   A. Baseline how long it takes to system test on similar projects. Assigned to Jon.

   B. Get some metrics on how many defects there have been on projects in the last year attributed to poor requirements/poor analysis (i.e., lack of time). Assigned to Mhadavi.

   C. Create a proposal for management. Assigned to Marybeth.

   D. Deliver the presentation to management. Assigned to Jason.

Regardless of how many tasks there are, there needs to be accountability. If your team wants to make a difference, they can all participate in the effort.

We find this technique to be easy and quick. It prevents people from complaining too much and creates accountability to implement the solution. This technique does not even require flip charts and sticky notes. It can just be a way of life. Get it out of your head (N). Get your heart into it (P). Make it a reality (T). Nip It!

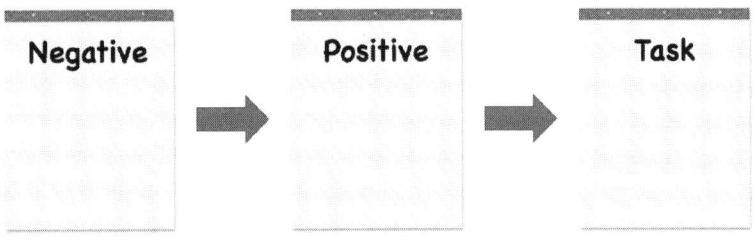

Prioritize Your Reflections

Reflection can lead to an overwhelming number of issues and failures. Although we have created actionable items with NiP iT, we must be thoughtful and prioritize our list of issues and failures. This is key to determining which issues or failure should be tackled first. Focus on the top 3-5 issues and failures and find ways to address them, so they will not be repeated in the next attempt. Put solutions in place to solve your top issues and failures.

Tackling the top issues and failures creates increased focus. Trying to "Boil the Ocean" by tackling a significant list of issues and failures is too difficult and time consuming to achieve. Handling a smaller list will allow you to focus on a few and produce better results. We want to fail fast. To fail fast, we need to focus. We want to keep moving quickly. If you are unable to resolve an issue or failure point, put it aside and pick up another one to try to solve. Keep your eye on the future. You may not be able to address all high priority issue for one reason or another but be flexible to refocus on another top issue or failure.

Framing Yourself as a Beginner

"Hey, new guy here." When someone first starts a job or project you can rest assured that those words will be thrown out to the group at some point. We do this as a way of setting group expectations of how you will interact with the group. Assuming the role of "new guy" tells the group, you are a learner or a beginner. The group is more patient and takes a little more time helping the new guy get the context or lay of the land during discussions. Groups treat new guys more patiently and are more open to explaining things in more detail to ensure understanding. The opposite is framing yourself as an expert. You are expected to know the answers to the situation immediately. You are after all the expert with a lot of experience. The group's expectations are very high for an expert, and they expect quick results. Even experts who have significant experience still need to frame themselves as a learned to change the dynamic, so the group is patiently explaining things more openly.

Being a consultant, you learn over a few engagements that framing yourself initially as a beginner can be done in many ways. "Every organization configures tools a little differently to meet their needs. Could you help me understand how your organization configured this tool?" What you are doing is framing yourself as a beginner in that organization. You still retain your expertise in the tool. Why is this of any benefit?

The benefit is that you have shifted the group's thinking into a more patient and explanative mindset. The group sees you as a beginner, so they are more open about the configuration of the tool. They won't assume you know everything about their configuration and this provides you with a better overall understanding.

Would a consultant walk in and frame themselves as an absolute expert? We would advise against framing yourself as an absolute expert. It's challenging for a group to work with someone who is framed as the absolute expert. We define the absolute expert as someone who has complete comprehensive knowledge. Does a group want to deal with the expert know-it-all? Probably not. Framing yourself as the absolute expert can create the impression the group's knowledge is not on equal footing with the expert, and the group's experiences are less important. The group then doesn't fully engage with the expert or holds back information because they believe the expert will figure it out on their own at some point. The group is alienated from the expert. In the end, the expert gets too little insight from the group and reaches an incomplete understanding that leads to drawing wrong conclusions. Even when you are an expert, there are advantages to the entire team for not acting like it all the time.

We all start off as beginners when we are presented new experiences throughout our careers and lives. Framing ourselves as new or inexperienced allows us to go into a learning mode more comfortably. It also sets the expectations for how a group should treat us as the learner.

When we have a significant amount of experience with something, we tend to primarily pull from our previous experiences and find it more difficult to move into a learning mode. Experiences can be limited and using them immediately is our brain's way to shortcutting the thinking to respond quickly. As human beings, we have a compelling need to react to situations, so we feel that we are in control and even to be viewed as intelligent. A fast response doesn't always make the best response.

The "New Guy" status can stick to you for years if you let it. The goal is to move into experimental learning mode and be less of an observational learner over time. In the beginning, observation is a good way to start learning a new service or product, and it's a great way to get the lay of the land when learning a new product or service. "Getting Up to Speed" means being able to move through observational learning into experimental learning quickly. It means getting your hands dirty quickly but not recklessly. Carefully take it all in and observe before trying to experiment. Moving into experimental learning forces you to drop that "New Guy" label.

Being a beginner sounds a lot easier on paper than reality. To be successful at learning, you will need to yourself as a beginner. Present yourself as a learner. "I'm here to learn more about this product. Can you tell me more about it?" This approach is a great way to introduce yourself and frame yourself as a learner.

Use your experience differently. Apply your experience to explore all the possibilities – not to limit them. Don't let your experience or current knowledge prevent you from exploring all the possibilities. Avoid putting on those rose-colored glasses when observing a new service or product you are learning. Be willing to explore with a new learner frame.

Be playful and curious when you are ready to experiment. This opens the door to looking at things in a new light. Remember to keep your experiments in a safe environment and avoid impacting customer's or the organization's reputation. Play fearlessly in a sandbox environment or test environment. Play and interact with a sample product. Spark your curiosity to learn more about it.

Seek out others who are outside of your group that are experienced in the product and ask them for help in understanding the product in greater detail. External experts can help you see things from a more global perspective. Reading the documentation isn't the only option to learn about a product. It's getting hands-on experience with an expert while interacting with the product that will give you a significantly higher level of understanding. Don't be shy about asking lots of questions to help you learn more. You're a beginner, and beginners ask lots of questions.

When learning a new system, keeping a list of important things you learned. Hands-on interaction will help you retain more knowledge about the product in the future, but don't feel that taking notes is a bad thing.

Failing Abundantly

Another part of framing failure is to understand is that to be great, you first must fail abundantly.

1. If I want to be a great musician, I must first play a lot of bad music. How do you get to Carnegie Hall? Practice. Well, GPS on your smartphone helps too.

2. If I want to be a great project manager or business analyst, I must work on a lot of projects that fail. Nobody wants to work on a project that fails, but we learn a significant amount of knowledge by being a part of a failed project.

3. If I want to learn French, I must speak a lot of bad French. By being corrected on pronunciation, I learn to speak French better. By not learning to get my point across in French, I learn how to phrase my point better so it's understood in French.

4. If I want to succeed, I must fail at least some of the time.

Some of the most successful people are those that have failed spectacularly. Let's talk catsup, or ketchup if you feel that word failed. Henry J. Heinz [5] didn't start out being a ketchup king. He started selling horseradish sauce. The company began in 1869, but due to the labor-intensive product and the financial crisis in 1873, the company filed for bankruptcy in 1875. One primary lesson they learned was

---

[5] Wikipedia, en.wikipedia.org/w/index.php?title=Heinz&oldid=859025220, Heinz, September 11, 2018

that Heinz customers preferred the product in clear bottles where they can see the product.

In 1876 Heinz opened his new company and introduced the world to Heinz bottled ketchup. Heinz became a massive condiment company with ketchup, relish, mustard and even horseradish. Heinz also expanded further into pickles and baked beans. Arguably Heinz learned a lot from his first attempt at being a horseradish company and applied it to make an even more significant success in his next company.

Certainly, this says something about believing in the power of failure. Despite bankruptcy, the following people bounced back from their lessons learned in failure to become very successful: [6] [7]

- H.J Heinz (Heinz)
- Henry Ford (Ford Motors)
- Walt Disney (Walt Disney)
- George Foreman (Boxer turned Grill Spokesman)
- Clarence Saunders (Piggly Wiggly now Safeway)
- Milton Hershey (Hershey products)
- William C. Durant (General Motors and Chevrolet)
- Ulysses S. Grant (President)
- Francis Ford Coppola (Movie Director)
- Stan Lee (Marvel Universe)
- Abraham Lincoln (President)
- Burt Reynolds (Actor)
- P.T. Barnum (Showman and Circus Owner)

---

[6] Incomediary.com, www.incomediary.com/went-bankrupt-now-worth-millions, Went Bankrupt, Now Worth Millions!, Michael Dunlop, August 1, 2018
[7] James Gentile Law Office, azlawyer.net/successful-people-filed-bankruptcy/, Successful People Who Filed for Bankruptcy, September 11, 2018

Failing and learning from the failure gives you the knowledge to make your next attempt at a project, company, or endeavor more likely to succeed.

Let's take learning Italian or some other language you have never spoken. What happens those first few times you learn to speak this new language? You are nowhere near speaking it like a native, and your instructor is making many corrections as you speak. Many mistakes are made in attempting the language, and you learn from those mistakes. After a large number of failures, you learn to speak this new language reasonably well. Well enough to say "Acqua, Bagno, Grazie, Ciao, Perdono, Si, No, Bene, Aiuto, Per Favore, and Arrivederci." In English that would be Water, bathroom, thank you, hello, I'm sorry, yes, no, great, help, please and goodbye.

To succeed you will need to fail abundantly. Find the lessons learned from failure and use them in your future attempts. Hopefully, it will not sound like this: "Ciao. Perdono, no grazie, no bene acqua bagno, aiuto si? Arrivederci." Translation: "Hello. I'm sorry, no thank you, no great water bathroom, please help yes? Goodbye."

Failure Isn't a Lack of Intelligence

When we fail, we are seen not as learners, but rather as troublemakers or lacking intelligence. We can also view ourselves as lacking intelligence when a failure occurs. We beat ourselves up mentally for failing. We need to frame failure as something that is needed to learn and that even the most intelligent people will fail.

George is walking through the park one day and sees people playing chess in the park. He finds the game of chess very interesting and remembers he had played chess when he was younger. George thought he wasn't very good at chess anymore and decided to brush up on chess skills. He bought a computer game to teach himself chess. George went home to practice on his own with the computer game. After a few weeks of practice, George was ranked as an intermediate level player by the computer game. At this point, George thought it was time to challenge other chess players. He decided it was time to take a trip to the park.

George went to the park and sat at a chess table. One of the chess players in the park asked George for a game, sat down, and started to play. George was incredibly focused, nervous and very competitive. His opponent was casual, carrying on multiple conversations and laughing with the players around him. His opponent barely noticed the chess board.

As the morning turned into an afternoon, George played chess with a few different opponents in the park. All his opponents were casual and carried on conversations with others around them. George never talked to anyone and was focused on the game ignoring everything else. George was losing game after game in the park as his opponents casually made unpredictable moves.

After the last game, George had enough. He got frustrated and determined that chess just wasn't for him. He quit chess that day because he didn't feel he was smart enough or talented enough to play.

There are a few lessons that we can all learn from this story.

Being playful and casual when learning makes the learning experience more exciting and less frustrating. Don't focus on being perfect or winning as this will close down your ability to learn. Learning in a negative environment is more difficult than learning in a positive environment.

Failing at something doesn't mean you lack intelligence, it means you are learning. We all questions our sanity or intelligence at some point, especially if we fail at the same thing over and over. Frame the failure as a learning opportunity.

Be vulnerable and understand that as a beginner or a learner you will need to keep trying and failing to learn. Sometimes taking a safe and reasonable chance on something new will produce incredible opportunities to learn.

Be open to learning from others. Learning on your own is not as productive as learning with others. As you learn from others, they learn from you.

What is Experiential Learning?

There are several different types of learning, and one we will focus on is experiential learning. Experiential learning is a fundamental concept behind the method for harnessing failure. Once you have failed, the next important step is to learn from the failure.

The definition of Experiential Learning[8] is the process of learning through experience and is more specifically defined as learning through reflection on doing. (Source: Wikipedia)

Experiential learning is a hands-on approach to learning. It's how we learned to crawl or walk. No instruction manual or book told us how to walk. We just had to experience walking to learn it.

The core of experiential learning is learning by doing. It is the belief that we learn best by physically doing something rather than just reading about it. Experiential learning theory typically focuses on the individual's learning through interaction. However, some of those principals of experimentation and learning from hands-on interaction can be seen in how organizations and teams learn. I would argue that organizations create learning by trial and error. Building a prototype or learning environment also creates a place where experimentation can be employed to learn. Beta testing by releasing software to select users for testing is also another approach.

Experiential learning has unique advantages when compared to other types of learning such as academic learning and participation learning.

Academic learning [9] is defined as the process of acquiring information through the study of a subject indirectly by lecture and reading. The primary difference is that the learner is not active, but

---

[8] Wikipedia, en.wikipedia.org/w/index.php?title=Experiential_learning&oldid=857287865, Experiential learning, August 30, 2018
[9] Wikipedia, en.wikipedia.org/w/index.php?title=Experiential_learning&oldid=857287865, Experiential learning, August 30, 2018

somewhat passive in the learning experience. The retention of what is learned is low.

Participation learning [10] is defined as performing exercises, discussions and demonstrations in a controlled environment. The primary characteristic of participation learning is that the learner does not interact with real-world models or situations. The retention of what is learned is low to moderate. (Source: Wikipedia)

Below is a diagram shows how experiential learning compares with styles or learning for retaining the information:

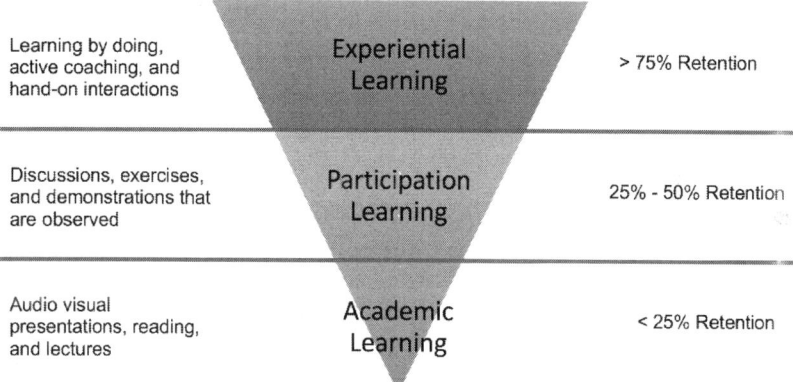

Retention of learning is vital to both the individual and organization as it allows the learner to apply what they have learned to new situations in the future. Building a deeper reservoir enables the individual or organization to pull from more experienced and

---

[10] Wikipedia, en.wikipedia.org/w/index.php?title=Informal_learning&oldid=856093224, Informal Learning, August 22, 2018

knowledgeable individuals. More enlightened individuals can address conditions faster with better results.

Many experiments have shown us that hands-on learning produces better learning results, but there are other factors in making experiential learning even more useful. Here are a few dimensions[11] that will assist in making experiential learning more useful for an individual or organization.

Self-Initiative. Experiential learning requires the desire or intention to learn. How can learning occur if the learner does not initiate the desire to learn? The same can be said for an organization. If the organization doesn't initiate or put forth the effort, they get zero learning. The initiative is the drive and desire to learn.

The stronger the desire or passion behind the desire to learn can be a significant driver in learning more and learning faster. In situations where you are required to learn something but have no passion for it, how much do you remember? Compare that to cases where you had a strong desire and passion. Where you more excited and engaged? Did you learn more? When individuals are excited about a new challenge, product or project they want to learn everything about it in great detail. If that passion lacks, then the desire to learn is not enough to complete the challenge, product, or project.

Active Learning. You need to be experiencing the learning while doing it in the actual environment. The requires the learner to

---

[11] Wikipedia, en.wikipedia.org/w/index.php?title=Experiential_learning&oldid=857287865, Experiential learning elements, August 1, 2018

interact with a physical model or product to learn it the most effectively. For a change of pace, there is no reading of documentation or books on the physical model or product. Active Learning occurs within an organization when those within the organization interact with models or products physically to learn.

Reflection. Once the interaction is completed with the physical model or product reflection can begin. Reflection is looking back at the experience of interacting physically with the challenge, product or project. Get feedback on the positive and negative aspects of the experience. Organizations have traditionally achieved this through product focus groups or other methods of consumer testing. Beta testing is another approach that gathers reflections on the product for further improvements.

Critical Thinking. Critical thinking is an objective analysis of data and facts to form a rational and unbiased conclusion or judgment. Being skeptical in your thinking prevents the analysis of data and facts from being biased toward a specific solution or outcome. Critical thinking is an important attribute to have when reflecting on results.

These dimensions (Self-Initiative, Active Learning, Reflection, and Critical Thinking) set the stage for the learner to be the most effective at experiential learning. These dimensions also form the Fail Fast Model.

We have now learned the definition of experiential learning, got a basic understanding of how it works, and understand the key dimensions in making experiential learning useful. However, how does this work in an organization or team?

In organizations, Experiential Learning is observed in internships, Gemba (a Six Sigma approach), observation of colleagues performing tasks, job-shadowing, cross training, product demonstrations, and in other areas. In these cases, there is Self-Initiative to learn either because of your desire or the influence of the organization on your willingness to learn. Individuals within the organization are given hands-on time to experiment and interact actively with the new product or tool.

Prototyping is an excellent way to engage in Experiential Learning. By giving individuals a physical product to interact with, individuals can learn more quickly and provide more in-depth feedback. Prototypes are very useful at eliciting customers or consumers' needs and desires. As interaction with the product occurs, customers or consumers can see potential uses in day to day activities.

Incorporating play time with training is often an effective tool for users to learn and provide feedback. By interacting with the product or tool freeform (without a specific list of tasks to complete) the customer or consumer goes into an exploration mode. They don't have a specific goal in mind but learn a lot of good stuff taking the journey. Training play time is very effective when you need individuals to support the product or tool. By playing with it - even trying to break it - they are learning more deeply and will be able to support it more effectively.

Learning Models

To fail fast, we must learn how to learn fast. To understand how to learn faster, let's look at two learning models. The first is Kolb's Experiential Learning Model (ELM) and the second is Bloom's Taxonomy. The models demonstrate how we learn and how we can learn more effectively.

Kolb's Experiential Learning Model Introduction

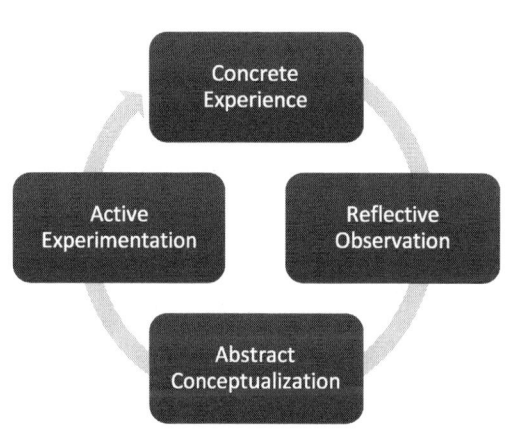

David Kolb's Experiential Learning Model (ELM) is a well-recognized model that explains experiential learning. Kolb's experiments along with other researchers experiments have proven ELM as a valuable model for understanding learning. Let's walk through ELM to define it better.

Concrete Experience

Concrete Experience is defined as having the learner performing a task or interacting with a product in the "here and now." The learner is physically and actively engaging in a hands-on manner with the task or product.

Concrete Experience is about understanding the product or system in its current state. Hands-on interactions, observations, and other techniques are used to fully understand and elaborate on the products features, capabilities, usability, fit for purpose, and other factors.

Concrete Experience is obtained by the organization when a team is interacting with a product in the "here and now." In technology terms, this is interacting with the product in the production environment. In business terms, it is interacting with the product in a real-world setting. The organization is physically and actively engaging in a hands-on manner with the product to create a concrete experience of the product.

A team gets a better experience when every single member of the team gets hands-on experience. Watching others interact with the product dulls the understanding and experience. Lack of interaction causes team members to produce limited amounts of reflection. It leads to team members being less familiar with the product and thus more inclined to not provide much reflection in the reflective observation phase.

Disney Parks have used the concrete experience phase of ELM to help their upper management to understand the operations of their theme parks better. Vice Presidents, Directors, and managers across the organization have spent time working in their parks. Disney's company leaders worked as any other park employee including wearing the uniform, interacting with park guests and even operating the rides. This concrete experience was crucial to improving their guest experience and products.

Reflective Observation

Reflection is a crucial part of the experiential learning process, and like experiential learning itself, it can be facilitated or independent. Reflection is a critical part of the Experiential Learning process. Reflective Observation can be facilitated, performed individually or a combination of both. The individual can reflect upon their personal experience to organize their thoughts before bringing them to a meeting where the organization reflects upon the failure as a group.

Reflection elicits the experiences of interacting with the product by individuals. Reflection creates the desired state for the product. Vision, goals, and objectives for the future of the product can be derived from Reflective Observation. Reflection also produces the dreams and hopes for the product that are not always based in reality.

Organizations reflect on concrete experiences in many different ways. A debrief meeting that reviews the entire experience, Agile retrospectives, lessons learned meetings and other approaches. The objective of the organization is to gain a 360-degree point of view. By eliciting the interaction experiences from all team members, the organization can build this 360-degree view of everyone's experiences with the product. Involving diverse individuals from around the organization can help achieve the goal of getting a good point of view. Eliciting from a single unit or department within an organization leads to narrow points of view which could distort the reflection on the experience with the product.

"Plus-Delta", "NiPiT", and "Good, Bad, & Learn More" are solid techniques to use in this phase. The key to being successful in this phase is to understand the experience in terms of things that went well, things that didn't go well, and things that were not understood or we didn't know why they were happening. Look at the list of things that are not understood. If that list is too long, reconsider going back to the concrete experience phase to gain more experience with the product.

Reflect on the good and not so good experiences you had with the product. Staying on the negatives side of things is easy. Keep the positive and negative in mind during the abstract conceptualization phase. Take the negative experiences and prioritize them. Focus on getting to the root cause for the highest priority negative experiences.

Abstract Conceptualization

Patterns emerge from Reflective Observation, and you begin to understand these patterns and how to harness a better approach to more successful conclusions of the experience. The learner is looking back at what they experienced (Reflective Observation) and based on that reflection of the discover new ways to improve (Abstract Conceptualization).

Conceptualization creates the future state. Conceptualization allows the individual or team to take reflections and turn them into actionable tasks for the product experience. Primarily this takes the vision, hopes, and desires to define those that are feasible or are aligned with the root cause of the negative experiences discovered during the Concrete Experience phase.

In this phase, the organization is taking the outcome of reflective observation to perform root-cause analysis to create a suggested plan of action to resolve the root cause of those issues discovered in Reflective Observation. The organization or team is looking back at what they experienced (Reflective Observation) and based on that reflection of the discover new ways to improve (Abstract Conceptualization) to improve. Experiments are designed to address the improvements and to prove out the root cause analysis is valid.

Abstract conceptualization works well with root cause techniques like the Ishikawa Diagram (Fishbone Diagram), or the 5 Whys. A big part of abstract conceptualizing is problem-solving and figuring out the approach that will best eliminate the negative experiences when interacting with a product.

Active Experimentation

Taking the Abstract Conceptualizations and creating experiments or environments in which Concrete Experience can be obtained. A safe environment is created in which failure and learning can occur.

This is taking the future state and actively changing or developing the product to create a new current state.

In this phase, the team is taking the outcome of abstract conceptualization to create specific tasks or experiments to improve the product. The team is actively making changes to the product to improve it. Once the agreed-upon changes are completed, the team will go into the concrete experience phase. Prototyping, conference room pilots, and BETA testing are effective at helping to define the improvements that remove negative experiences.

Experiential Learning Model Example

Let's walk through an example of the Experiential Learning Model (ELM). For this example, you decide you want to learn more about how a dairy farm works after reading an article online about dairy farmers. You chose to learn more and will go to a dairy farm to get the first-hand experience of being a dairy farmer. Your self-Initiative started your desire for learning. Your desire to learn starts you down the path to gain concrete experience.

You interact with the farmer having conversations about his farm and how it works. You make observations watching the farmer care for his dairy cows and learn about the schedule the farmer must maintain to keep his dairy farm running. You ask to help the farmer milk the cows to see how the milking process works. Here you are Active Learning. You are physically interacting with the farmer and the cows to learn more about dairy farming. You are in the concrete experience phase.

After you have finished observing the dairy farmer and the dairy cows for a day you return home and reflect on your experience. You share your experience with your family and draw conclusions based on this reflection. By looking back at your experiences over the day and drawing conclusions, you are in reflective observation.

Based on your experiences you see ways in which you could interact with the dairy cows better. Taking your experience and building new ways to interact with the dairy cows is abstract conceptualization.

Building a plan with specific tasks you would like to try again with the techniques you feel will improve the interaction with the dairy cows is active experimentation.

The DMAIC Connection

Now that we have walked through ELM comparisons to the Lean Six Sigma methodology of DMAIC [12] (Define Measure Analyze Improve and Control) are coming into the minds of Lean Six Sigma Black Belts. Arguably, ELM is an underlying set of principles that are expanded on in the DMAIC model.

Let's understand and define DMAIC. DMAIC is a data-centric model for stabilizing products, improving products, business processes, and product designs. DMAIC is commonly linked to Six Sigma but is used for many purposes.

ELM is a model that shows how you learn from your experiences. Using ELM will allow you to learn and deeply understand the product.

ELM defines how you learn. DMAIC defines how to improve. That is a subtle difference between the models. Both models play well with each other. Learning is a key part of both models.

---

[12] Wikipedia, en.wikipedia.org/w/index.php?title=DMAIC&oldid=857375261, DMAIC, August 31, 2018

We will discuss ELM and Agile in future chapters of this book to demonstrate how ELM assists you in learning, but how to use that learning to improve faster.

Bloom's Taxonomy

Bloom's Taxonomy[13] was created in 1956 by an educational psychologist Benjamin Bloom to understand the process of learning in both gathering knowledge and applying that knowledge. In 2001 Bloom's Taxonomy was revised by the team of Anderson and Krathwohl in response to several criticisms of the taxonomy. Educators follow Bloom's Taxonomy because it helps to develop learning objectives that help explain the process of learning.

One of the outcomes of the research is a pyramid model which shows how we first start to gain knowledge and then move up the pyramid in using that knowledge to develop new concepts, ideas, or products.

Understanding Bloom's Taxonomy helps you understand how you take the knowledge that is acquired from concrete experience to deepen your understanding and application of the knowledge. Let's walk through Bloom's Taxonomy and discuss how it assists in learning to fail fast.

Do you remember when you first learned to tie your shoes? Well it did not start out with your parents saying, "Tie your shoes,

---

[13] Wikipedia, en.wikipedia.org/w/index.php?title=Bloom%27s_taxonomy&oldid=858745740, Bloom's taxonomy, September 9, 2018

Madeline." It started with "This is a shoe." "This is a shoelace." "This is your foot." You had to learn the names of things and then you learned why you needed them. "A shoe is to protect your foot." "A shoelace keeps the shoe on your foot." Then you learned how to do it. "This is how you tie your shoelaces." Practice, practice, practice. Next, you would understand why the shoe was not staying on because the shoelaces became untied or your friends untied them when you were not looking! After you decided to ditch those so-called-friends, and possibly look at better show lace options, you finally make new friends, and your new shoes with Velcro makes your life easier.

Let's dig a little deeper into Bloom's taxonomy to determine how the application of the model will help us understand how to fail fast and learn quickly.

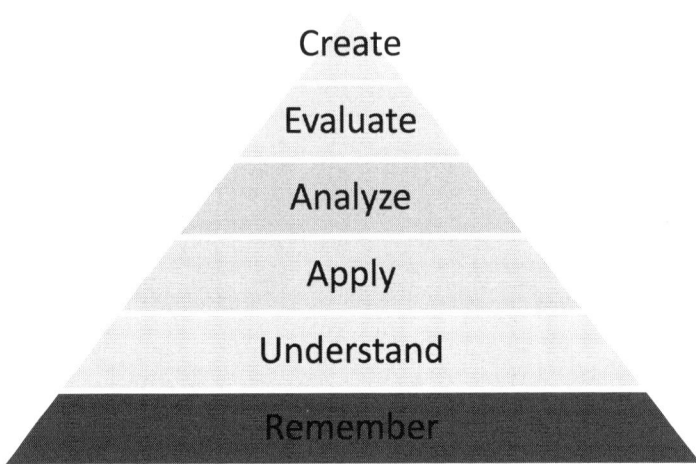

Remember

Recall or retrieve previously learned information. Example: reciting rules or policies from memory.

Recalling your concrete experience (either good or bad) is an essential step in gaining knowledge and being able to apply that knowledge. Taking notes and being hands-on with the product helps you to remember experiences with it more clearly. Passively observing can help gain insight, but actively engaging in the experience and applying knowledge that was previously gained is beneficial for learning faster.

If you had experience with a forklift previously, then your adventure with a brand-new forklift allows you to use previous knowledge to learn the new forklift faster. However, it can also be a deterrent to learning if you believe you don't need the experience because all forklifts act in the same manner. This generalization will cause you not thoroughly learn about the new forklift and instead rely on old knowledge of a previous forklift that might be out of date. Generalization can create biased thinking. Avoid biased thinking.

Approach the learning openly and get a hands-on experience with the new forklift for an unbiased experience.

Understand

Understanding is comprehending the meaning of a problem. Being able to state a problem using your own words. Example: Explaining a complicated process in your own words and understanding the rationale of the process.

As you gain knowledge of the forklift's operations with hands-on experience, you begin to have both positive and negative experiences with the forklift. At this point, you are taking new knowledge and previous experiences (if you had any past experiences) and applying them to understand your positive and negative experiences more fully. This allows you to more fully state the positive and negative experiences in your own words. You know the rationale of the forklift operation and why some things that appear as negative are in fact positive. As an example, the safety feature of a loud alarm if you are not sitting on the seat of the forklift correctly.

Apply

Use what you have learned in a new and unprompted way. Taking what you have read or observed and applying it to a unique situation. Example: Reading the employee manual and applying the vacation policy to your upcoming trip to Europe.

The forklift has given you a negative experience – an example would be that it didn't start. Taking your previous knowledge and your understanding of the operation of the NEW forklift, you seek to understand why the situation has occurred. You might review the troubleshooting tips in the manual and apply those tips to your current situation in an attempt to resolve the problem. You remember observing others interacting with the forklift and use those observations to solve the problem. You don't fully understand how all the parts fit together and work on this forklift, but you know it enough to execute steps to try to fix the forklift.

Analyze

The ability to troubleshoot and understand how something is put together by being able to determine the difference between facts versus guesses. The ability to gather information from others, digest that information and communicate with others. Example: Troubleshooting a printer to determine why it is not printing as expected.

Back to the forklift. You have gathered enough knowledge on the forklift and how the parts of the forklift work together. Perhaps you took it apart and put it back together, replaced a part, or performed some maintenance activity. Based on this level of knowledge you can determine potential areas or parts of the forklift that are causing the failure to occur. You can perform troubleshooting tips and logically go from part to part to find the problem.

Evaluate

Make judgments about the values of ideas around the product or problem. Example: Deductive reasoning and logical thinking to select the best possible choice based on pros and cons. The ability to justify the selection of a solution.

At this level, you can hear a noise or see a movement that allows you to diagnose an issue with the forklift with reasonable accuracy. You know all the parts of the forklift and have encountered a few negative experiences with their operation. This leads you to find the broken part more quickly. You can make snap judgments about the forklift's operation problem. The engine didn't start when you turned the key, you didn't see the panel light up, and understood the red indicator light on the dashboard. Knowing what was wrong, you were able to make the best possible choices for fixing the problem or failure. You plug it in and charge the battery on the forklift. You can justify why you believe the battery is dead based upon your experience and knowledge.

Create

Building new structures or patterns where they did not exist before. Developing new capabilities for products. Taking existing parts and recombining them into a new product. Example: Taking existing components and designing a new product or system.

With all your forklift knowledge, you are now able to design a whole new generation of hydrogen forklifts that run on water, not a battery. You can deconstruct the forklift and retrofit the forklift with a new hydrogen engine. You are taking existing parts, creating new parts, and assembling a never seen before forklift.

Bloom's Taxonomy is important to consider as you go through the ELM (Experiential Learning Model). As you move through the ELM cycles, you are continuing to learn more by stepping up – or stepping down – Bloom's taxonomy model. Failure – and success – are the drivers that inspire you to learn. Understanding that failure is a critical component of learning allows you to accept that failure and move forward in the learning process.

Let's explore Bloom's Taxonomy in a way that is very relatable to our day to day activities.

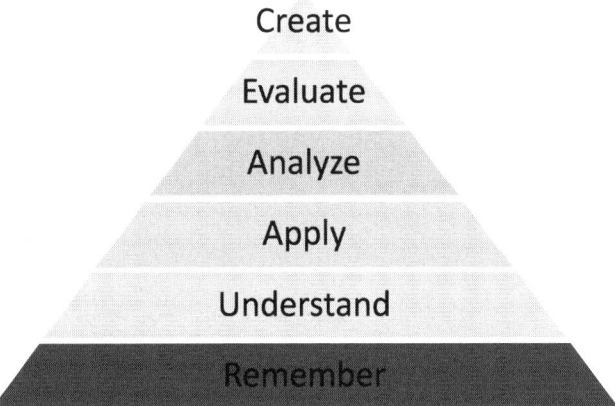

People jump to the analyze level of the pyramid above, skipping the remember, understand and apply levels without really knowing what they are doing. Example: How do you define a customer? How many times have you seen a lack of shared understanding of definitions that sends people off into two different directions? So many levels were skipped that the results obtained on the analysis level were incomplete and doomed.

We were listening to a story of a student in a class who was told to go to the store and get "Chitlins" for dinner. Sounds easy right? Just look for them in the store. After a visit to the store, the student couldn't find the Chitlins and returned home empty-handed. A family member then pointed out that Chitlins were in the frozen section. Back the student went to the store with a member of the family. In the frozen section they found a bucket named "Chitterlings," paid for it and went home. Not only did the student not know the right term, but they also did not understand anything about Chitlins. Just sounds like something fancy for dinner. Chitterlings or Chitlins are the small intestines from a pig. Are you going to eat them?

In this case, it was a great lesson to learn and failing was not that big of a deal. An extra trip to the grocery store, some frustration, and total loss of appetite. It did help the student to learn that she needed to start on Bloom's remember phase first. If she did not know the definition, then she needed to ask and learn. Of course, upon discovering the true meaning, she would not have gone. It is also why the family member went the second time. Fearing that she would not bring them home when she figured out the definition of Chitlins.

What if this were on a multi-million-dollar project? The failure would be massive. Using the principles of Bloom's Taxonomy in your life or organization will make your failures more palatable and acceptable. Failure because we experimented is one thing – failure because we did not understand a noun is another. Know your terms, and what they are for, then you can choose the right experiment to fail.

Critical Thinking and Failure

Critical thinking is a significant part of failing fast and failing safe. When engaging leaders to talk about training needs, we often hear the response "We need our people to be better critical thinkers." Well, here we go. Our first response is to ask them "How do you define critical thinking?" The initial response we often get is "We need them to do their jobs better." We start performing root cause analysis, and we eventually get to the response, "We need them to make better decisions." That gives us insight into what the organization's leaders are truly expecting. Better decision making in an organization is essential. We seem to be onto something here, but what is critical thinking you ask? Read the following seven definitions. Which one is the right definition of critical thinking?

1. The objective analysis and evaluation of an issue to form a judgment.

2. Critical thinking is the intellectually disciplined process of actively and skillfully conceptualizing, applying, analyzing, synthesizing, or evaluating information gathered from, or generated by, observation, experience, reflection, reasoning, or communication, as a guide to belief and action. In its exemplary form, it is based on universal intellectual values that transcend subject matter divisions: clarity, accuracy, precision, consistency, relevance, sound evidence, good reasons, depth, breadth, and fairness.

3. Disciplined thinking that is clear, rational, open-minded, and informed by evidence.

4. Critical thinking is a way of deciding whether a claim is true, partially true, or false. Critical thinking is a process that leads to skills that can be learned, mastered and used. Critical thinking is a tool by which one can come about reasoned conclusions based on a reasoned process. This process

incorporates passion and creativity but guides it with discipline, practicality and common sense.

5. Critical thinking is a process that begins with an argument and progresses toward evaluation. Three interrelated activities activate the process:

    a. Asking essential questions designed to identify and assess what is being said.

    b. Answering those questions by focusing on their impact on stated inferences.

    c. Displaying the desire to deploy critical questions.

6. Thought processes that are quick, accurate, and assumption-free. Such processes help us view, with a critical eye, the problems, decisions, and situations that require appropriate reaction and action.

7. The ability to make decisions, solve problems, and take action as appropriate.

Which one is correct? All of them are correct. They are simply different versions with the same intent. The one you chose as correct is likely the style or approach you are going to take with critical thinking. Some will choose a more complex view of critical thinking and others may choose a simple statement to reflect how they see it. You would both be correct.

Critical thinking separates fact from fiction, fallacies in reasoning, and removes bias to make better decisions. Critical thinking relies on several right (soft) and left (hard) brain approaches to thinking. A balanced approach between the right brain and left brain will yield better results that help you make better judgments. You see our brains break things down into the hard and soft. Left and right. Logic vs. Creative.

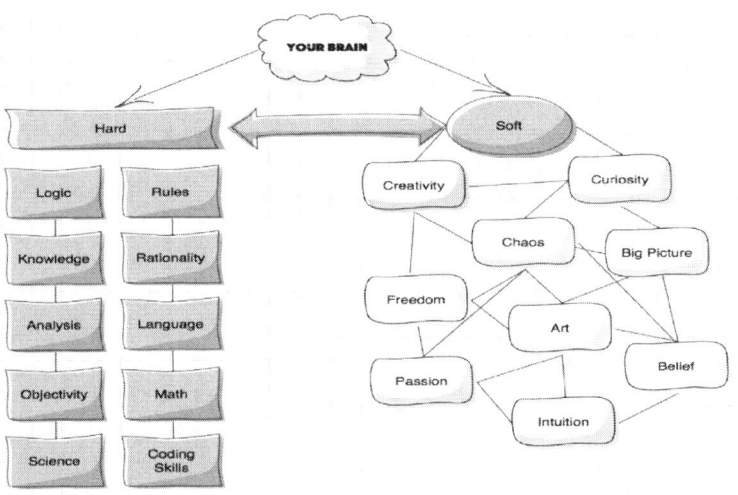

You make judgments when you are experimenting. These judgments will determine the size, speed, and impact of your failure. Remember, the point of this book is to fail fast and fail safe, so we need to understand the types of critical thinking that will help us fail fast and fail safe.

If you only use right-brained, soft/creative approaches your failures might take longer, or they will be more significant - a size you cannot recover from financially, emotionally, or will create the CLM (career limiting move).

If you only use left-brained, hard approaches your failures might not yield unusual or exciting results which equals a lack of innovation.

10 Ways to Inspire Your Team to Think Critically

For ages, employers have preferred workers who can complete their tasks and follow whatever instructions their bosses give. Doubt and pushback were viewed as problems by managers who wanted employees to tow the party line. Today, however, startups and Fortune 500 companies alike are turning to doubters who question every decision within the company.

These critical thinkers help poke holes in company plans to strengthen them and provide alternative solutions that are better in the long run. Keep reading to learn how you can foster a team of critical thinkers to move your own company forward.

Why Companies Value Critical Thinkers

Big data has given companies more information than they could ever hope to process. While analytics tools help soft this information, it's up to strategic business analysts to ask the right questions and assemble the data in a way that's helpful. Today's employees aren't number crunchers who spend hours balancing spreadsheets; they're strategists who manage the technology presented to them.

"When brawn was valued over brain, workers were asked to check their critical thinking at the door," Ira Wolfe writes at Talent Economy[14]. "[Today], they're expected to carry out a growing list of

---

[14] Talent Economy, www.talenteconomy.io/2017/01/31/6-critical-skills-tomorrows-workplace/, 6 Critical Skills, Ira Wolfe, January 2017

responsibilities and meet higher expectations with fewer resources. That requires them to think analytically and apply the results of all that thinking without the benefit of supervision or experience."

Instead of completing a set of tasks each day, employees in the modern workforce are constantly challenged to complete more tasks or complete the same tasks with less. This has made the economy less of a production line and more of a puzzle.

Not only is critical thinking an essential job skill, but it's also a tool for managers who want to streamline their departments. Kadie Regan at Filtered [15] highlights a few ways critical thinking can help in the workplace. Two of the top benefits include saving time and enhanced communication:

1. **Time**: You already know that not all information is relevant to decision-making, and critical thinking allows you to remove unnecessary information to get to the heart of the problem.

2. **Communication**: By focusing only on relevant information, you can communicate your thoughts more effectively and persuade your audiences.

Managers can quantify how much creative thinking helps their business and improves their processes.

---

[15] FilterRed, learn.filtered.com/blog/6-benefits-of-critical-thinking, 6 Benefits of Critical Thinking, Kadie Regan, August 2015

Improving your ability to plan strategically saves your department time and makes everyone better informed. It also helps your team in other ways. Essentially, critical thinking is a gateway skill. Breanne Harris at Pearson[16] writes that people who score well in critical thinking assessments are also rated highly by their supervisors in:

- analysis and problem-solving,
- decision-making,
- overall job performance,
- creativity,
- job knowledge,
- the ability to evaluate the quality of information presented,
- moreover, the potential to move up in the organization.

Further, the Brandon Hall Group and Laci Loew[17] surveyed more than 2,100 managers on what they thought would be the most important skills required of leaders in the next five years. Thirty-five percent said critical thinking was the most important, followed by collaboration (32 percent) and creativity (28 percent).

How to Spot Critical Thinking in Job Candidates

When talking to employees and potential job candidates, you can't just come out and ask whether they're critical thinkers. Most candidates would self-identify as critical thinkers. Instead, Career

---

[16] Pearson, www.pearsoned.com/education-blog/the-status-of-critical-thinking-in-the-workplace/, The Status of Critical Thinking in the Workplace, Breanne Harris, September 2015

[17] Brandon Hall Group, http://www.brandonhall.com/blogs/critical-thinking-the-difference-between-good-and-great-leaders/, Laci Loew, July 2014

Metis[18] recommends looking for specific personality traits that set critical thinkers apart from the crowd:

- Self-confidence in their ability to reason.
- Curiosity and the desire to remain well-informed.
- Flexibility in considering multiple options.
- Transparency about personal biases.

Critical thinking requires an employee to form their own decisions and to understand that their decisions might change when new information comes to light. One of the best ways to test critical thinking is with examples.

Dr. John Sullivan[19] offers a few ways companies can screen for critical thinkers during the hiring process. Consider presenting a real scenario that the candidate might encounter and ask how they would solve it. Ask follow-up questions to get them to defend it or present new information that makes them change their answer on the fly.

Dr. Sullivan also suggests presenting a flawed strategic plan and asking the candidate to identify potential issues within the process. This highlights a candidate's ability to question plans and come up with alternative ideas.

---

[18] Career Metis, www.careermetis.com/benefits-critical-thinking-workplace/, The Importance And Benefits Of Critical Thinking and Reflection In The Workplace, Alice Jones, November 2016

[19] Harvard Business Review, https://hbr.org/2016/12/6-ways-to-screen-job-candidates-for-strategic-thinking, John Sullivan, December 2016

10 Actionable Steps to Develop Critical Thinkers

Critical thinking is one of the most universally in-demand skills in the workforce today. Samantha Cole [20] reports at Fast Company that 72 percent of employers think critical thinking is important to an organization's success, but only half of employers said their employees demonstrate this skill.

This leaves employers with two options: Fire existing employees or train them to become critical thinkers. Here are a few steps you can take should you choose the latter.

1. Give Employees Ownership

"All too frequently, we hear managers complain that their employees don't think for themselves," Francesca Gino and Brad Staats write at the Harvard Business Review[21]. "Yet these same managers punish their subordinates for failing to follow instructions. To avoid this pitfall, managers need to separate the outcome from the process. That means specifying what good work should look like without locking down all elements of the process."

Instead of providing a set of instructions for each project, consider explaining what an ideal finished product would look like. You can also provide resources for help if your employees get stuck. This

---

[20] Fast Company, www.fastcompany.com/3037837/employers-want-critical-thinkers-but-do-they-know-what-it-means, Employers Want "Critical Thinkers, Samantha Cole, October 2014

[21] Harvard Business Review, hbr.org/2015/06/developing-employees-who-think-for-themselves, Developing Employees Who Think for Themselves, Francesca Gino & Bradley Staats, June 2015

means they have to connect the dots to make something happen instead of following your orders.

2. Encourage Discussion and Debate

Give decision-making power to your employees by asking them to propose multiple solutions to a problem. Executive coach Joel Garfinkle[22] encourages managers to debate their employees on each point, highlighting the pros and cons while asking them to defend their points of view.

Not only does this help employees recognize their own biases, but it also prepares them to defend their ideas when they need to make similar pitches to upper management.

3. Adopt 'Orphan Problems'

"In every organization, there are annoying problems no one claims as their own," Art Petty writes at The Balance[23]. "Identify an orphan problem and ask for your boss's support in tackling it."

Orphan problems give your team the opportunity to think critically to solve a problem outside of their normal job duties. These challenges will also give them a chance to enhance their skill sets and get noticed by management. Solving a problem that no one wants to

---

[22] LinkedIn, www.linkedin.com/pulse/4-ways-train-employees-critical-thinkers-joel-garfinkle, 4 Ways to Train Employees to Be Critical Thinkers, Joel Garfinkle, February 2015
[23] The Balance, https://www.thebalancecareers.com/strengthen-critical-thinking-skills-2275911, 6 Exercises to Strengthen Your Critical Thinking Skills, Art Petty, August 2018

own shows initiative and leadership that can pay off at promotion time.

4. Create Risk-Taking Opportunities

Failure is an important part of any career. To help your employees think critically, let them develop their own solutions and execute the ones they think are best. They may fail, but this gives them an opportunity to find solutions and learn from their mistakes.

DeLynn Senna[24] writes that the best managers give employees the chance to take risks. If your team adopts an orphan problem that no one wants, this could be a safe opportunity to try something new. The risk factor is lower than a revenue-critical project but still, lets your employees try something new and pick themselves up if something goes wrong.

5. Set Aside Time for Reflection

In an article for the Harvard Business Review, Nina Bowman[25] writes that some people are so focused on accomplishing a set of tasks each day that they don't have time for critical thinking.

"[One coworker] kept a jam-packed schedule, running from meeting to meeting," Bowman writes. "She found it difficult to contribute strategically without the time to reflect on the issues and to ponder options."

---

[24] Robert Half, https://www.roberthalf.com/blog/salaries-and-skills/discover-how-to-boost-your-strategic-thinking-skills, Discover How to Boost Your Strategic-Thinking Skills, DeLynn Senna, July 2016
[25] Harvard Business Review, https://hbr.org/2016/12/4-ways-to-improve-your-strategic-thinking-skills, 4 Ways to Improve Your Strategic Thinking Skills, Nina A. Bowman, December 2016

By delegating tasks and eliminating unnecessary meetings, you create more time to focus on problems and to identify creative solutions. This way, you can be actively engaged in meetings, not just a warm body that got an invite.

6. Reward Thinking Instead of Reacting

Some modern offices operate in a constant state of crisis. Whenever a problem arises, employees rush to put out the fire as fast as they can. This can lead to short-term solutions that create bigger problems in the long run.

Robert Kabacoff[26] encourages managers to reward employees who demonstrate that they've spent time thinking about a problem and considering their options. The best critical thinkers will identify multiple solutions and work through the one that has the best long-term benefits, not just a short-term plug.

7. Add Speed to Your Problem-Solving Meetings

This might seem to counter the advice to think instead of react, but forcing employees to think on their feet fuels creativity and forces employees to get out of their comfort zones.

Maren Hogan[27], CEO of Red Branch Media, starts meetings with a fast brainstorm to write down as many solutions to a problem as possible within the first few minutes. Everyone is encouraged to list a

---

[26] Inc, www.inc.com/will-yakowicz/how-to-foster-strategic-thinking-in-employees.html, How to Get Your Employees to Think Strategically, Robert Kabacoff, February 2014

[27] Poll Everywhere, blog.polleverywhere.com/creativity-during-meetings/, 12 unusual ways to spur creativity during meetings, Maren Hogan, August 2018

few options and creative solutions. Then, the team narrows the most plausible and lists the pros and cons of each. This also creates a team environment for employees to support each other's ideas and defend their own.

8. Communicate the Company's Big Picture

"Understanding the broader organizational strategy helps employees incorporate it into their work," Tawny Lees[28] writes at CEO.com.

Not only will your employees start to tie their ideas to the company's goals, but you can use overall corporate goals to brainstorm possible ways to improve your department. Furthermore, you can identify the ways other teams have come up with goal-aligned ideas to use as examples for your team to follow.

9. Foster a Question-Friendly Environment

Two of the most dangerous phrases within a company are "because I said so" and "because we've always done it that way." Both of those phrases close the door on learning why something is done and identifying ways to do something better.

Todd Wallis[29] encourages employees to create a question-friendly work environment for curious minds. By teaching employees to ask open-ended questions without one right answer, they can start to

---

[28] CEO, www.ceo.com/leadership/5-ways-to-develop-more-strategic-thinking, 5 Ways To Develop More Strategic Thinking, Tawny Lees, August 2018
[29] Vail Valley Partnership, www.vailvalleypartnership.com/2017/05/critical-thinking/, The Need for Critical Thinking in Your Organization, Todd Wallis, May 2017

evaluate their choices and the processes around them for continuous improvement.

10. Use Both Sides of the Brain

Critical thinking requires the employee to use both sides of the brain: They need to find solutions and logically defend them creatively. Pam Warren[30] encourages managers to create more creative opportunities for employees in highly logical or analytical positions, and vice versa. When both parts of the brain go through mental workouts, they're likely to work well together when the time comes.

Critical Thinking Terms

Did you know 22 types of critical thinking can assist with your failing fast and failing safe?

Let's test your knowledge of the 22 different types of critical thinking. Match the critical thinking type with the letter. See the list below. Match the letter from the definitions list below to each one of the critical thinking terms in the box provided.

For example, the critical thinking type of Analytic is best described by definition A.

This test isn't easy and will take some effort to complete it. Our goal here is not to write a book about critical thinking. We intend to

---

[30] LinkedIn, www.linkedin.com/pulse/how-improve-strategic-thinking-skills-leadership-pam-warren, How To Improve Strategic Thinking Skills In Leadership?, Pam Warren, February 2016

provide an understanding of the different critical thinking types. Give it some thought as some of the answers will make more sense as you think through them. This test will help you understand these thinking types clearly. We hope you will give a shot.

If you desire not to take the test, we won't hold it against you. Skip to the answer page and read each critical thinking type and its definition.

Critical Thinking Type	Definition	Critical Thinking Type	Definition	Critical Thinking Type	Definition
Analytic		Associative		Creative	
Systems		Divergent		Focused	
Generative		Intuitive		Linear	
Yes… and		Yes… but		Convergent	
Diffuse		Lateral		Probability	
Objective		Possibility		Subjective	
Critical		Verbal		Visual	
Vertical					

Definitions

A. The abstract separation of a whole into its constituent parts in order to study the parts and their relations.

B. Thinking that moves away in diverging directions to involve a variety of aspects and which sometimes lead to novel ideas and solutions; associated with creativity.

C. Proceeding from or taking place in a person's mind rather than the external world.

D. A logical approach to thinking – the opposite of lateral thinking.

E. The desired result a person or a system envisions, plans and commits to achieve a personal or organizational desired end-point in some sort of assumed development.

F. The mental process of making associations between a given subject and all present pertinent factors without drawing on experience.

G. Solving problems through an indirect and creative approach, using the reasoning that is not immediately obvious and involving ideas that may not be obtainable by using only traditional step-by-step logic.

H. Uses the part of the brain that is emotional and creative, to organize information intuitively and simultaneously.

I. It is the understanding and reasoning using concepts framed in words. It aims at evaluating the ability to think constructively, rather than at simple fluency or vocabulary recognition.

J. A way of looking at problems or situations from a fresh perspective that suggests unorthodox solutions (which may look or feel unsettling at first).

K. An exhaustive approach where the inclusion of all events occurs, or an exclusive approach where multiple events do not occur simultaneously.

L. The agreement or understanding through speech to make or express rationale or a decision.

M. Using sense-perception only as a starting point, to bring forth ideas, images, possibilities, ways out of a blocked situation, by a process that is mostly unconscious.

N. A term used to describe the process of finding a single best solution to a problem.

O. The process of agreement with exceptions.

P. The objective analysis and evaluation of an issue to form a judgment.

Q. A process of thought following known cycles or step-by-step progression where a response to a step must be elicited before another step is taken.

R. A different view of organizations. Organizations do not travel a straight line and rational course from vision to mission to goals to strategy to execution.

S. A right-brained visual approach to thinking.

T. An approach to finding maximum clarity or distinctness.

U. Defining thoughts and feelings of what could be.

V. A holistic approach to analysis that focuses on the way that a system's constituent parts interrelate and how systems work over time and within the context of larger systems.

Spoilers and Answers

Critical Thinking Type	Definition	Critical Thinking Type	Definition	Critical Thinking Type	Definition
Analytic	A	Associative	F	Creative	J
Systems	V	Divergent	B	Focused	T
Generative	E	Intuitive	M	Linear	Q
Yes… and	L	Yes… but	O	Convergent	N
Diffuse	S	Lateral	G	Probability	K
Objective	R	Possibility	K	Subjective	C
Critical	P	Verbal	D	Visual	H
Vertical	D				

Take a few moments to review the answers above. Which critical thinking types do you use the most often? Which ones should you start using more when you are failing fast and failing safe?

We All Critically Think Differently

When people meet Bob, they often think physically - a big man, a big personality, a big risk taker, and a BIG change advocate. When they meet Paul, they often feel the opposite. Physically smaller, quieter personality, not a risk taker and cautiously for a change. We are Penn and Teller as it were. In reality, Paul is a bigger risk taker than Bob. He is more prone to do it because he is more comfortable with failing fast as long as he is producing a result. Any result. Bob had perfection beat into him for 20 years in corporate, Fortune 50 environments. Yes, 50 and not 500. Bob wanted to ensure you didn't think it was a typo. Beat it right into him. Although Paul is a bigger risk taker, Bob is perfectly fine with more experimentation and using more failure to get to the desired result. Paul takes less time in experimentation and drives it out faster.

So, who is right? Who is the better critical thinker? Who is better at failing fast and safe? Both of us of course! We are both using/bringing different parts of critical thinking to the table to make the necessary judgments and achieve the results that make sense to us. Could both of us do it better? Sure. We are learning every day from each other. From the outside, people could listen to us and think that we have the biggest arguments on the planet. Are we? No. We are passionately failing fast and failing safe while our critical thinking is at odds. We communicate a lot to get the result that we want, and we help each other fail in ways that the other would initially not be comfortable with.

Critical Thinking Quiz

Understanding how you ask questions or analyze information will lead you to judgments that will help you achieve your desired results. Let's put that to the test! If you have done this quiz before, we made changes to mix it up.

This 'critical thinking' test is typically used for young children as an introduction to critical thinking. According to Accenture (formerly Anderson Consulting Worldwide), around 90% of professionals get all the questions wrong. Why is that? As adults, we have formed bad habits, and we bring our personal baggage to the table that is biased based on our experiences, our value systems, race, creed, orientation, and more. Our bias gets in the way of critically thinking. If you get some questions wrong, don't worry about it. Perhaps it is time we start looking at things like our children do. With wonder and curiosity!

Critical Thinking Quiz

1. How do you put a giraffe into a refrigerator?

2. How do you put an elephant into a refrigerator?

3. The king of the lions is hosting an animal conference. All the animals attend except one. Which animal does not attend?

4. There is a river you must cross but it is used by crocodiles, and you do not have a boat. How do you manage it?

5. In the following line of letters, cross out six letters so that the remaining letters, without altering their sequence, will spell an English word.

    B S A I N X L E A T N T E A R S

6. What should the following actually say?

    VRYFINXMP

    LARXCDSW

    HATWXPCT

Spoilers and Answers

Question 1: How do you put a giraffe into a refrigerator? What did you say? Be honest; I have heard them all. Here are some of the answers I have heard over the years:

1. Get a bigger refrigerator.
2. Get two refrigerators and combine them to fit around the giraffe. Cut a hole in the top, so his head sticks out.
3. Why do you have to put him in a refrigerator? I would not do that. That sounds mean.
4. Ask a contortionist how they fit into a refrigerator.
5. Get a chainsaw, chop him up, and put him in.
6. Take him over to a warehouse where they have a huge cooling area, so he can move around and stay cool.

At this point, I start responding to answers which yield very interesting conversations:

1. Why do you have to get a bigger fridge?
2. Why are you solutioning?
3. Why do you think you do not have to do that?
4. A contortionist? Really?
5. Chainsaw? My you are practical (laughter ensues).
6. Warehouse? The question did say a refrigerator.

The answer to #1 is to "Open the refrigerator door, put him in and close the door." What? This question challenges our ability to not jump to solutions. We tend to go too complicated too quickly. If you want to fail fast and safe, don't jump to the solution so quickly.

Question 2: How do you put an elephant into a refrigerator? Here are some of the answers I have received:

1. Go out and buy a bigger refrigerator than the one you bought for the giraffe.
2. Going to need a bigger chainsaw!
3. Why are we talking about giraffes, elephants, and refrigerators? I thought this was about critical thinking!
4. Buy a stuffed animal elephant and put him in the fridge.
5. How can I answer this when I don't know how big the fridge is?

My responses to these questions are:

1. How do you know you have money to buy the fridge?
2. Bigger chainsaw? You have been watching too many horror films!
3. If critical thinking were easy, we would not need this test.
4. Aww…so sweet! Wrong but sweet.
5. Why do you have to know how big it is? Lots of debate assured on that question!

The answer to #2 is "Open the refrigerator door, take out the giraffe, put the elephant in and close the door." This question tests your ability to think through the repercussions of your previous actions. So many things we do are connected, and we need to see those connections and come to logical conclusions.

Question 3: The king of the lions is hosting an animal conference. All the animals attend except one. Which animal does not attend? Here are some of the answers I have received:

1. Hyenas
2. Scar
3. Insects
4. King of the lions
5. The elephant!

My responses to the answers are:

1. So, you are going with the hyenas from the Lion King? Because they are mean?
2. Yep. We have a Lion King movie theme going on.
3. Insects? Are they an animal? One?
4. Help me understand because the king of the lions is at the conference, right?
5. Are you sure? A triumphant smile crosses their face because they remember the answer to the last question.

The answer to #3 is "The elephant." You just put the elephant in the fridge. This question tests memory and how you use it. Did you make the association with the other questions?

Question 4: There is a river you must cross but it is used by crocodiles, and you do not have a boat. How do you manage it? Here are some of the answers I have received:

1. Build a bridge.
2. Take a long way around.
3. Build a boat and then cross.
4. I jump on the alligators backs one at a time like Indiana Jones!
5. I don't. I am risk averse, and I will just miss the conference. They can send me the notes.

My responses to the answers are:

1. So you assume you have materials? Unlimited time? I think the conference is about to start.
2. Do you have a special map that I do not have?
3. You and the bridge builder are going to be great friends!
4. Kudos for being Indiana Jones though. Have you done anything like that before? Gymnast? Track and field?
5. How do animals take notes?

The answer to question #4 is jump in the water and swim across. All of the crocodiles are at the animal conference. This question tests whether we can learn quickly from our mistakes. Are you starting to see the application of critical thinking and failing fast – fail safe?

Question 5: In the following line of letters, cross out six letters so that the remaining letters, without altering their sequence, will spell an English word.

B S A I N X L E A T N T E A R S

Here are some of the answers I have received:
1. Six letters.
2. Sales.
3. No clue.
4. Ban anteaters.
5. Banana.

I find there will be a few people that will get this one correct. The correct answer is **BANANA**. Cross out S, I, X, L, E, T, T, E, R, S.

B S̶ A I̶ N X̶ L̶ E̶ A T̶ N T̶ E̶ A R̶ S̶

This teaches us that not everything is connected (no connection to the first four questions) and that some things are literal. S, I, X vs. 6.

Question 6: What should the following actually say?

VRYFINXMP

LARXCDSW

HATWXPCT

Here are some of the answers I have received:

None - ever. It is extremely difficult. What some people will figure out is that an "E" is needed to solve the puzzle but because it uses words some would call old-fashioned, it is harder to solve.

Here is the answer:

Every fine exemplar exceeds what we expect.

The first time we did this test we failed too. It took a significant amount of effort to change our thinking to solve the puzzle.

Whether you got some questions right on this test, or you didn't get a single question right doesn't matter, but it does prove out the point that we need to work on our critical thinking skills. We make better decisions when we start gaining long-term mastery of our critical thinking skills. The purpose of the test is to give you a few things to consider now that it has exposed you to some essential types of critical thinking.

Different Thinking Challenge

Here is one more challenge that we are confident you will answer most or all of the questions correctly. Hint: there is one type of critical thinking that will help solve all of these questions.

1. Someone falls out of a thirty-story building, but lives. With luck and their landing pad not being factors, how could they have survived the fall?

2. There are a dozen eggs in a carton. Twelve people each take a single egg, but there is one egg left in the carton. How?

3. A boat has a ladder that's ten feet long, and hangs off the side of the boat, with its last two feet submerged in water. If the ocean tide rises five feet, how much of the ladder will be underwater?

4. There are ten birds perched on a fence. A farmer aims his rifle and shoots one. How many birds are left?

5. What weighs more – 100 pounds of feathers, or 100 pounds of quarters?

6. Which countries have the 4th of July out of the United States, the United Kingdom, France, Germany, Australia, and Canada?

7. If you were alone in a dark cabin, with only one match and a lamp, a fireplace, and a candle to choose from, which would you light first?

8. Does the Z go above or below the line?

<u>A   EF   HI   KLMN   T   VWXY</u>
   BCD   G   J   OPQRS   U

Different Thinking Challenge Quiz

1. They fell out of the ground floor. If you recognize that a floor was not specified, you will come to this conclusion. If not, you will invent (solution) the answer to include a landing pad of some kind.

2. The last person took the carton with the last egg. When you remove one egg at a time, you will eventually get down to just one egg. The scenario does not say the person must take the egg out.

3. Two feet. The ocean rises and so does the boat. If you used the right form of critical thinking you would realize this was not about math but a misconstruction of the question.

4. One. Just the dead bird the farmer shot. It is easy to imagine the farmer shot all 10 birds with one trick shot or other plausibly stretched reasoning. The farmer fires a single shot, and the other birds fly away.

5. They weigh the same. If you are not observant and you get distracted by physicality, you will calculate the weight based on a feather or the metal. A feather is lighter than metal.

6. All countries have a 4th of July. Don't get fooled by metadata – July 4th is Independence Day only in America. No other country has a July 4th holiday.

7. You light the match first. You can't light anything else until you have fire.

8. The "Z" goes on the top line. Did you state that it is because all of the letters on the top line have straight line parts? All of the letters on the bottom line have curved line parts.

Final Thoughts on Critical Thinking

Here are some final thoughts are critical thinking (because there are 100,000 things we could say):

- Be a self-critic. Constantly evaluate what you are doing and why.

- Don't be afraid to ask the questions especially questioning assumptions.

- Reverse your thinking. It will lead to more intelligent judgments to fail fast and safe.

- Remember that critically thinking 100% of the time is not going to happen. Critical thinking is a tool, and we don't always use the tools we have. Habits!

- Analyze the group influences around you. They shape judgments all the time.

- Redefine how you see things. You must critically think about yourself just as much as you critically think about the effort you are working on.

Creating Failure Habits

Are you ready to embrace failing fast and failing safe? Let's do a little test. Cross your arms like you always cross them. You feel comfortable crossing your arms because you have crossed your arms that same way every time. It is almost like hugging a friend. Now cross your arms the other way or the opposite of the way you typically cross your arms. If your right arm is usually on top, then put your left arm on top. Oh, awkward? Did you get it? I love watching people trying to cross their arms in the opposite direction. Imagine a room full of 600 people trying to cross their arms differently. It starts to look like a disco dance. It doesn't feel quite right. Why is that? We have consistently crossed our arms one way throughout our lives. We developed habits and these habits that have formed over many, many years. So, if I tell you to get used to change, get used to failing you should be able to do it right away? Not quite. Your habits will get in the way. So how long does it take to break or make a habit?

If you have read Stephen Covey, it takes 21 days. Twenty-one days comes from a book popular in the 60's called Psycho-Cybernetics by Maxwell Maltz. Maxwell was a plastic surgeon who noticed his patients would take about 21 days to get used to their new features. Well hold on a minute, that is very different from weight loss, building systems, and failing at work. So how long does it take?

Researchers from University College London published a study in The European Journal of Social Psychology that examined 96 people. They found the average time to develop a new habit that sticks is 66 days. Additionally, the study found that an individual's habits will require between 18 and 254 days! Bob knows this to be true given that he is a bigger man (girth and stature). Bob has lost the same 50 lbs. Every year for about ten years in a row. Not only is that bad for his health but it is terrible for forming habits. Bob needs to break the cycle. Bob never gets beyond the 254-day mark. Bob has to think differently about how his approach.

Telling our employees to fail fast without your employees understanding what it will take to be successful at failing is a fail in itself. To fail quickly, we need to break old habits and learn to fail fast. Not telling your management what it will take to be successful at failing is a fail in itself (and also not the intent of this book). Take this section from the book and have a talk about failure in your organization. Talk about what it will take to develop good habits around failing. Be open. Be transparent. Understand its importance.

Here are some tips for developing habits around failing:

1. Use Bloom's Taxonomy to have conversations. Remember the structure? By iterating through the model, you will find you develop good habits of communication around failure.

2. Make room for feelings and address them. "UGH!" said several people in their heads after reading that statement. There is psychology happening here with failure and feelings are a big part of it. Think more about being a global bartender and not a therapist.

3. Remember to communicate that anything meaningful has a risk and therefore you have to keep trying.

4. Stop playing the failure games, blamestorming, finger pointing, self-loathing, or other games.

5. Focus on the importance and lessons learned from the failure and not the person.

6. Always KISS. Keep It Simple Stupid. Don't run off and scream "Fail now! Fail today! Fail at everything!" Try to change one thing today and potentially more as you develop your failure skills.

7. The failure buddy system. Buddies can keep us accountable. Buddies help us develop habits.

8. Remember to use 'but' and 'and' a lot. When you immediately start with a negative, immediately follow up with a positive. "This is terrible, but we have an opportunity to make it better."

9. Do more retrospectives (lessons learned). You may not like them, but that is how we keep up with failing and learning.

10. Use visualization to assist in breaking or forming a habit. Example: You are about to do something you should not like eating that triple scoop chocolate covered ice cream cone. Visualize yourself picking it up, then putting it down, clap your hands together like you are washing your hands-free of the situation, then laugh and run away! Now do it a few more times in your head and start developing a good habit.

11. Make it public! It will make you and those you work with accountable.

12. Stick with it. Persistence. Perseverance. Stick-to-itiveness. Patience. Durability. Preservation. Maintain. Survivor mode.

Permanence. Tenacity. Indefatigability. Stamina. Well, you get the point.

Acceptance of Failure

One of the toughest parts of failure is accepting that you are not entirely in control of everything around you, but you are in control of how you react to things out of your control.

"It's not what happens to you, but how you react to it that matters."

- Epictetus

Epictetus was a Greek philosopher that believed we have control over our creativity and ideas but have no control over much else. To Epictetus, all external events are beyond our control, and we should calmly accept whatever happens. This philosophy has influenced Roman Emperors, writers, philosophers, and the US military.

When we experience failure, we suffer from a lot of negative emotions. We get frustrated, angry, or disillusioned by our failure. Failure happens all the time around us. We react to these failure events in different ways. Some we choose to ignore and others we get very frustrated. Failure happens either way. Our reaction to failure is what changes.

Part of failing fast is accepting that failure happens and sometimes it's mostly due to something entirely out of our control. By taking failure in a favorable light – as something to learn from – we are better able to harness failure to learn from it.

Even as we calmly accept failure ourselves, there will be others around us that will not. It's important to recognize that others may see your failure as negative. The negative can be made positive by showing others that this failure as given us the opportunity to learn and grown. Because the failure occurred, we are stronger and more knowledgeable.

We need to permit ourselves to fail. It's okay to fail at something because you will have the ability to learn from it. Recognize that without failure and the negative experience around it, we might never have the opportunity to learn and grow.

Several military leaders have quoted Epictetus and his philosophy as one of the primary reasons they were able to survive as prisoners of war enduring torture for significant periods of time. The acceptance of working through things that were not in their control most likely played a large part in their survival.

It's about the story you tell yourself. When you experience something, you can choose to tell that story as negative or positive or even somewhere in between. An effective way of handling failure is to change the story you tell yourself. Instead of "I messed that up!" change the story you tell yourself to "That was a failure, but this is an opportunity for me to learn and grow by reflecting on it honestly."

The story we tell sometimes does not reflect the reality of the failure. We need to objectively reflect on the failure and be honest with ourselves to learn and grow.

Before starting an experience that might result in failure, change your thinking. "I know this might not work out perfectly, but this is an opportunity to learn and grow." Frame your mind into working with the failure in a more positive light. We are not advocating jumping out of a plane without a parachute. Take reasonable risks for failure. In the fail safe section of this book, we talk about how to plan for a soft landing.

Once your expectations are set, it is time to set the expectations of others. Let them know where potential failures could occur. Set the expectation that experimenting and failing is a meaningful way to increase the organization's ability to learn and grow quickly. Provide the context that we can learn from failure in a significant way. Be sure to set the stage that these are reasonable risks of failure and that you are not proposing something entirely off base. It can take a lot of time to build trust with your team around failure. Let them know it's okay to fail and how to best learn from it. You will build confidence with your team with this approach.

Quantity versus Quality

Let's experiment with a pottery class to show how permitting yourself to fail is a strength and why abundantly failing is productive. For this experiment[31][32][33], a classroom filled with students who had little or no experience with pottery making were divided into two groups.

---

[31] Art and Fear, Ted Orland and David Waylon, April 2001
[32] Fail Fast Fail Often, Ryan Babineaux and John Krumboltz, December 2013
[33] Excellent Journey, excellentjourney.net/2015/03/04/art-fear-the-ceramics-class-and-quantity-before-quality, Art & Fear, Eric Johnson, March 4, 2015

The first group was called the QUANTITY team. The team was told that if they created more than 50 pounds of clay pots as a team, they would get an A for the class. Also, the QUANTITY team was also told there were no limits to the number of clay pots they could make. Then the team was informed that at the end of the class all pots created by both teams would be judged and a winner chosen by a panel of experts for technical merit and artistic quality.

The second group was the QUALITY team. The team was told that as a group they must make one perfect clay pot that had exceptional technical and artistic merit as judged by the instructor. The team was also instructed that at the end of the class all pots created by both teams would be judged and a winner chosen by a panel of experts for technical merit and artistic quality.

Both groups were given three months to finish their assignments. At the end of the three months, which group do you think was rated the highest for technical and artistic merit by a panel of pottery experts?

The QUANTITY team Won!

The QUANTITY team won [34] because instead of trying to avoid mistakes or achieving design perfection, they merely went about experimenting. Since there were no limits other than to make more than 50 pounds of clay pots, the team could experiment with many techniques. The team pushed themselves to make more pots and in doing so learned more. More experimentation was leading them to better designs which lead to higher technical and artistic merit.

---

[34] Excellent Journey, excellentjourney.net/2013/08/27/quantity-leads-to-quality/, Quantity leads to quality, Eric Johnson, August 27, 2013

The QUALITY team of students focused on being flawless to produce just one perfect pot and didn't make many pots. By focusing on creating only one ideal clay pot, they didn't frequently experiment and instead concentrated on design perfection.

In the end, the QUANTITY team gave themselves permission to fail and experiment. The permission to fail gave the team a different mindset and allowed the team to learn from failure.

Do you give yourself and your team permission to fail so they can learn? How could this be applied to your yourself or a team? Not giving yourself or your team permission to fail creates the situation where everyone is so focused on the quality that he or she never experiment, see the outcome and to innovate.

Organizational Permission to Fail

When you think of an organization giving permission to fail, you think of approval from your boss. As individuals within an organization, we can permit ourselves to fail. We do this by playing in non-production or safe environments where our experiments will not impact our customers.

A big part of granting permission to fail in an organization is that the organization must understand the value of experimentation and failure as a way for the organization to learn. Most organizational leaders have the understanding that you must have real-world experience with the product or service to be effective in maintaining and innovating a product or solution. Seek out opportunities to directly interact with the product or solution in the real world.

One of the most amazing experiences I had when working for a start-up medical device company was going to a hospital to observe our product in a live procedure with a patient, technicians, nurses, and doctors. Being in the accounting department, we were pretty removed from what our product looked like or even its capabilities. After that experience, we had a more profound understanding of the importance of the medical device we manufactured. We also learned about the significance of lot number tracking, managing consignment inventory, and processes that would help those in the operating room interact with our product more efficiently.

When we returned to work, we took the experience and experimented in our test systems to help us devise new ways to serve our customer's needs more effectively. We permitted ourselves to try out new ideas to see the outcomes. We failed a lot but kept at it. The result of permitting ourselves to fail and to experiment was that it resulted in a brand-new approach to meeting our customer's expectations. When we presented the improvements, our customer's response was, "Wow! You guys get how we work!"

We have worked for companies that are the direct opposite. These organizations believe they are invincible and utterly immune to a failure of any kind.

If you always think you will never fail, you never see failure coming. You just aren't looking for it. The invincible attitude of "we've never failed, we don't even know what failure is because we have always succeeded" will blind side an organization every time. Instead of embracing the opportunity to fail as a way to learn, the organization spreads a false belief that it can never fail. These organizations get very comfortable with success, and failure becomes a distant memory.

Organizational invincibility has a lot to do with the Emperor's New Clothes game we talked about at the beginning of the book. It a game that is often played in two variations:

- We can't fail – we're invincible!
- We can't admit failure because we don't want to face the fact we failed.

Invincibility is a pretty dangerous game for organizations play. By not learning from their failure, they merely repeat it over and over. The same mistakes and outcomes happen in an endless loop.

We consulted with a company like this where a specific project was attempted and kept failing. The project failed, but no one stopped to figure out the lessons that could be learned from the failure. The failure was swept under the rug and forgotten. Then the organization's leadership would start the project up again. Attempts to bring in lessons learned into planning were ignored. The organization didn't learn from past experiences and didn't want to learn from the past failures. The reason? "That team didn't know what they were doing. We have a better team now!"

True invincibility is the ability to fail and learn from it. Rapidly learning from failure propels the organization forward at fantastic speeds. Speed so fast that it appears you are invincible. It is not until you put everything in slow motion that you see the failures, and the lessons learned from those failures were applied immediately.

One of the biggest mistakes for an individual or organization to make is not taking the lessons they learned and applying them back to improving the product or service. Just gathering up lessons learned isn't enough. Reflect on lessons learned. Find the cause of the failures and make improvements in processes. You need to take the action of incorporating into your environments and experiments. Live the ELM Model. Experience it first hand, reflect on that experience, get to the root of the problem, and implement changes that will remove the source of the problem.

When failure does hit the invincible organization, they don't know how to handle it. The organization stumbles around and doesn't recover quickly from that failure. They never learned how to fail fast and adapt to improve quickly.

Chapter 3

Curiosity and Failure

"I have no special talents. I am only passionately curious."

- Albert Einstein

Easily recognized as one of the 20th century's leading scientists and thinkers, Einstein is a classic example of curiosity and its power as a tool for experiential learning. We all see Einstein's intelligence, but do we see his curiosity? Curiosity is a critical part of failure.

An excellent example of curiosity being a motivator or initiator in learning is observing younger children. Younger children experience the world with all of their senses by pushing out and exploring. It seems their energy to study can be boundless. They are always fidgeting, poking, looking, asking questions and experimenting to learn more about the world around them.

Children in general rarely think about the impact of their exploration of the world around them. They are fearless about learning new things and certainly don't stop and think about how their action will impact their near future or even future careers.

Curiosity is an integral part of our adult lives as well. Through curiosity, we continue to learn as adults. As we grow older and wiser, we become a little more careful and at times fearful of situations which can take away our passion for learning. As the old expression goes, "When you stop learning you die." It's a bit harsh, but there is some truth in those words as well. Being ambivalent or refusing to learn leads to having the number of our experiences diminished. So, what happened? Your curiosity has waned.

Without learning experiences, we lose our ability to apply those lessons learned from those experiences and apply them to future situations or issues. It is akin to putting yourself into a bubble whereby nothing new ever happens. It's rather boring and doesn't generate much for brain activity. Who wants to be boring? Did anyone raise their hand? Don't let your curiosity get killed off like a character of a TV show.

Curiosity Killers

As adults, we have fears that can prevent us from being curious. These fears can get in the way of us wanting to explore and be curious. We don't want to look like children or foolish, we are expected to be experts, so we stop exploring our curiosity and deprive ourselves of the learning opportunities that curiosity presents.

Adults have an overpowering desire to see the world through "rose-colored" glasses. We all have a strong world-view which we look through when making our day to day decisions. Our lenses see the world in one way and allow us to refuse to challenge it or being curious about it. You will not be able to see the curiosity and opportunities that are around us for learning and exploring until you remove those "rose-colored" glasses.

Children don't have these lenses. Watching young children play for only a few moments and you begin to see that they have no fear. Let's make a mud pie and eat it! Ah, maybe not! Children have very high energy in learning about the world around them. As we grow into adults that energy fades. Bob does an exercise that asks students who ate paste as a child or glued their hands together and then peeled the glue off. There are always a few. Maybe not as many paste eaters but definitely hand gluers. They are extremely curious people.

There is a happy place in-between these two sides. A more central place that allows for curiosity to be sparked but isn't fearful of curiosity and learning. Failing fast but safely failing resides there. It's a great place to be. Many experts say that continually learning new things gives you stronger mental abilities and resiliency in bouncing back from difficult situations. The key is to find your happy place where curiosity can thrive in an adult world.

"Whether you think you can, or think you can't – you're right."
- Henry Ford

Have you ever been curious about something but then just decided to pass on the opportunity even though you never really tried it out? Killing your curiosity to provide a learning experience is easily done. At other times, we talk ourselves out of a new experience entirely without attempting an experiment. It's just too risky. It's just too silly. I don't have enough time. I'll probably be bad at it. Well, I guess you were right if you gave in to fear.

Indeed, there are things we are not interested in and would not bother to pursue, but why do we ignore the spark of curiosity when faced with something interesting? The spark of curiosity could be telling you there is something new and exciting to learn. An opportunity to gain knowledge is at hand.

Time is very important for us all. Everyone values time. We can use time as a weapon for us to kill curiosity and our ability to learn. Arguably, if you are passionate and interested in a topic, then you could make time for it.

We also kill our sense of curiosity by not wanting to look stupid. No adult wants to appear to be unintelligent or untalented in front of others. Experimenting with a topic doesn't' require you publicly succeed or fail at it. There is no law that experiments must be public. Attempting to make bread at home from scratch doesn't require a press conference. Today my fellow Americans I'm making bread! You can comfortably experiment with making bread in your own home.

Are you curious? Ask yourself the following questions:
1. Do you seek out new training opportunities?
2. Do you like meeting new people?
3. Do you like the excitement and exploration of the unknown?
4. Do you question rules because some of them are made to be broken?
5. Do you try new things just to see what they are like?
6. Do you enjoy trying new approaches to solving problems?
7. Do you welcome new ideas?
8. Do you challenge the status quo and are willing to be a disruptor?

If you can answer yes to most of these questions, your curiosity is alive and thriving! If you did not, it is time for a curiosity boost! To continuously be curious sometimes you just need some support.

If you are looking for support in experimenting, turn towards your spouse, trusted colleagues, or friends to get their insight and to encourage you to learn something new and exciting. They can also be there to give you support when the experiment fails or doesn't work out and can help you make a judgment call on trying that experiment again.

Do I still want to learn more about this topic that has sparked my curiosity? Should I continue to experiment? These are some great questions to ask yourself after your first few attempts - even if those attempts were successful or not. A successful outcome of the experiment makes answering these questions much more comfortable.

If the experiment is a disaster, then answering the questions above on whether to continue experimenting can be more challenging to answer. Be honest with yourself. Remember you are a beginner and a learner on this topic. Attempts at becoming a world champion bread maker overnight are not realistic. No champion was ever made by trying something just once or twice. Deciding to abandon learning after the first attempt can be giving up too quickly on something that you are interested in or sparks your curiosity. If your curiosity has faded away, it is probably time to stop experimenting and move on to learning the next topic that sparks your curiosity.

Organizations think in these same ways as well. Should they commit to it? Is it worth it? Organizations can learn by experimentation in many ways that can lead to better outcomes on products. Creating a safe environment to try out an idea that might work is critical to harnessing curiosity to experiment and learn.

Something frequently heard in organizations is "It costs too much!" If you love flying, you don't need to buy an airline to pursue your love of flying. Cheap experiments are a good thing and valued, but we must recognize some experiments will be costly. Being passionate about curiosity can help you prioritize your personal funding towards experimenting.

Organizations fear the outcomes of experiments with products. Creating a safe environment where those experiments can occur will help keep curiosity alive and kicking in the organization. Set the expectation a safe environment is being set up so that failure will occur – such as a proof of concept model. Set the expectations that the decision to continue with the product or product enhancement is dependent upon the outcome of experiments in the safe environment.

Curious Experimentation Drives Innovation

Organizations in general place little value on experimentation. Experimentation is seen as costly, wasting time and just playing around for the sake of playing around. Everyone talks about wanting to be like Amazon, Apple, Google, Microsoft. Well, those organizations see it differently. They see it as an opportunity for the organization's products to grow and improve. It's also a great way to keep employees engaged and happier. They did not grow because they did the status quo. They grew because they were curious and experimented.

"Curiosity, especially intellectual inquisitiveness, is what separates the truly alive from those who are merely going through the motions."
- Tom Robbins

Let's look at the Google organization and its products. Google has a policy with its developers that provides them with paid time to pursue ideas of their own and to experiment. Google created great ideas in its labs by developers experimenting. Not all these products made it out of the lab but think about what would have happened if they had not experimented. Would any of their excellent ideas have made the light of day? These experiments have driven Google into browser-based software, mobile operating systems, and even self-driving cars. All these products were far away from their core search and advertising business.

Harnessing curiosity for an organization can be exceptionally powerful if encouraged. Failure in experimentation can happen frequently, but it is through this failure that the organization and the individual both learn.

Stop Taking Yourself Out of Exploring

Open the door to curiosity. Let your organization attempt experiments that are not directly related to one of your organization's products. Curiosity brings out passion even when experiments fail. Harnessing that failure creates learning experiences. As individuals try out topics or activities, they often find other items that are interesting and pursue them. Experimentation sparks new ideas and innovation. High levels of experimentation change individuals into smart problem solvers, explorers and bring new capabilities that otherwise would be overlooked. What is learned from this curiosity is unknown.

Find a way to experiment with new topics to use your curiosity. You don't need to have elaborate and expensive experiments to learn. Small-scale and low-cost experiments are great ways to learn quickly and explore curiosity.

Pursue whatever you find interesting. A colleague of mine loved working with algorithms and equations to create solutions for reporting. Her passion for algorithms led her into the field of business intelligence and data science. With her assistance, her organization found new patterns in their customer's use of their product which lead to strong sales growth. Her passion on the subject led her to learn about data patterns and data reservoirs. She had no idea that one day her career would be to use her love for math and algorithms to solve real-world business problems. You don't know where curiosity will take you – until you get there.

A Story of Curious Exploration

When Paul was younger, he always found art to be exciting. Paul was interested in creating and designing. So, he was happy to go to art class. Art class, on the other hand, wasn't so excited. His first assignment was to sculpt with clay – sculpting anything at all. He got one lump of clay and tried working with it. As hard as he tried the only thing he could make was an ashtray which looked more like a flat square box. Paul's art teacher gave him a D. She was generous with that grade as even Paul didn't think it was worth that grade. The next assignment in art class was to create an oil painting featuring a still life. Anything with fruit in a bowl would be beautiful.

So off Paul went to draw his of a bowl of fruit. He sketched it repeatedly on paper and couldn't get it right. Finally, he decided to place items on the canvas generally and go from there. The bowl of fruit will be about there. A table was obviously underneath the bowl of fruit. An open window with a curtain in the back looked good. Paul painted the canvas over and over. He kept feeling he wasn't getting it quite right. Everyone only got one canvas, so he just kept painting over things to try to make it better. After many attempts, he finally got to the point where he couldn't paint anymore. It's not just going to get better he told himself.

Paul learned a lot about art in that class. He pursued something he found interesting. Paul took a significant chance of getting a lot of ridicule by turning in my assignment. Well, that bowl of fruit wasn't Picasso, but it landed him an A.

Take a chance and explore a topic you are passionate about to see where it leads you. Will it lead you to the misshapen ashtray or the painting of still life?

Advantages of Curiosity

Being curious also provides excellent benefits. By utilizing curiosity frequently, it will keep you highly aware and in the present. It can improve your ability to see a wider variety of options and open your thinking to solve problems faster.

Curiosity drives you to learn faster because your passion is pushing your desire to learn more about the topic. When you are passionate about a subject, you want to learn more about it. Passion and interest create positive energy which feeds your curiosity even more.

The energy created by curiosity drives innovation and invention. This energy feeds on itself to grow bigger and stronger. The more opportunities to be curious and experiment lead organizations to faster results for all the organization's objectives and goals.

The Curiosity List

Inspire yourself to be curious and try new things by creating a curiosity list. In its basic form, it is a list of things that you would like to explore further. Our curiosity list, bucket list, and reading list have all morphed together into one list. Here are a few things from our lists:

1. What is a rare earth magnet? Why does it stick to my granite countertop?
2. Understand Tesla's innovation thinking - read "Wizard: The Life and Times of Nichola Tesla."
3. Scrum Book of Knowledge: How does strategic enterprise analysis fit with it?
4. How would we learn simple conversational French for our trip to Montreal?
5. Do a survey of 100 people and see how many think the sun is yellow and polar bears are white, then gauge their reaction when they find out neither is true. Further critical thinking research by publishing the results.

It does at times look like a to-do list. The difference with a curiosity list is that we have chosen to think and further pursue a deeper understanding of these items. Take out the garage, clean the kitchen, get an oil change for the car or file receipts with your medical savings account aren't things that bring greater depth to your knowledge.

Understanding a rare earth magnet isn't critical to my everyday living, but it does enhance my ability to use that knowledge for other purposes. While exploring that concept online Paul discovered another cool idea of flexible cloth type magnets. These magnets are woven into fabrics and can be laundered. By placing the woven magnets in the collars, clothes could be hung by sticking them to the wall. Since the magnet was in the collar of the shirts, they all hung quite quickly on the wall. This story was very inspiring, and we were curious to see how we could magnetize a wall in the closet. Paul added this one to his list a few weeks ago and is trying to figure out a way to buy some materials to build a small-scale prototype. Anything to make laundry easier.

Put things on your curiosity list that you always wanted to try but never could either make the time. Look through the list and prioritize it to pick the top 3 to 5 items. Find a simple and easy way to explore items on the list. This will help you gain a better understanding of whether or not you want to pursue a full-scale implementation of your idea.

Want to learn transactional French for an upcoming trip? Figure out how you would accomplish it. Are there free courses online? Can you buy a book? Do you have a friend that speaks French? Is there an abbreviated quick way to learn transactional French?

Get your team or family hooked on the idea of experimenting. Try things out together and support each other as experiments are conducted. Build a support system that will allow you to explore even further. One of our colleagues has 4 children (5 if you include her husband, okay make that 6 as they are all kids at heart) and they are the most hands-on experiment-driven family I have ever seen. They create the most interesting and exciting things. Their kids are amazing – all 6 of them.

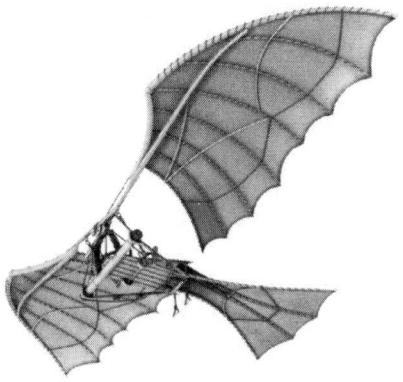

If all of this does not get you curious, you can always think about Leonardo da Vinci. He lived his life by seven principles: Curiosita (curiosity), Dimonstrazione (testing and experimentation), Sensazione (using your senses), Sfumato (being open to ambiguity), Arte/Scienza (using whole-brain (right and left) thinking, Corporalita (fitness/poise), and Connessione (interconnectedness – systems thinking).

When you look at the principles he lived by, many of them are embraced by the Fail Fast Fail Safe way of life. Leo, (as Bob likes to call him because in his mind they are on a first name basis) was so curious that nothing would get in the way of the curiosity. He would not just be curious about why the bird could fly, his curiosity would ask about all of the potential factors of bird flight.

- How is the bird structured so that it can fly?
- How are birds differently structured and yet still fly?
- What about the feathers and the formation of them aids in flight?
- What is the speed of wind that increases or inhibits flight?
- What is the weight a bird can be and still fly?

- How would we recreate that if many were to fly?

Curiosity. Whether you are an Einstein or da Vinci fan, they both had something in common. Curiosity. Curiosity that allowed them to visualize the future. Then they failed a lot to see what would happen.

Isn't it time for you to get curious and see what your failure can do for you and your organization?

Chapter 4

Challenging Your Organization to Work with Failure

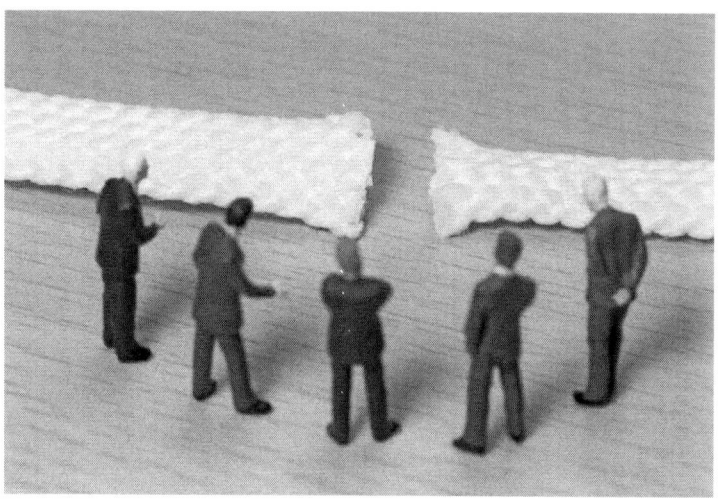

"Success is the ability to go from failure to failure without losing your enthusiasm."

- Winston Churchill

A culture change needs to occur as many organizations don't embrace the value of failure and failing fast. Let's look at how to change this organizational culture. Here are some key starting points to discuss with your organization on failing faster:

- Failure isn't a bad thing, and we learn from failure.

- Talk about ELM (Experiential Learning Model) and talk about how failure helps to learn.

- Discuss Bloom's Taxonomy and how we take knowledge and build it into creativity by building team members knowledge.

- Relate ceremonies like agile retrospectives and why they help us learn but only if we apply them to the next sprint.

- Continuously improve products but also the development and implementation processes that deliver those products.

Let's take a look at some ways to take these talking points and present them to an organization.

Convince the Boss (or Colleagues) to Try Your Big Idea

Moving your great ideas forward requires approval and buy-in from management. This can either be frustrating or terrifying, depending on your level of confidence. And even when know your proposed change would benefit the company and grow your department, you're aware that risk-averse teams might shut you down in favor of staying in their comfort zone.

With the right tools, however, it's possible to push upper management out of their comfort zones. Here's how anyone — from entry-level employees to VPs — can get buy-ins from their bosses to create change.

Persuading Management with Confidence

It's perfectly normal to get nervous talking to executives. They're often in a hurry, and it can be intimidating to ask for change or additional resources.

However, that doesn't mean you can't pitch your ideas confidently.

Brendan Reid[35], a career and management coach, compares many employees to wet noodles when they talk to management: They're so concerned about being respectful and following the organizational chart that they bend to the point of not getting what they want. Furthermore, they let their views change quickly if they think it matches what their managers want.

Reid encourages employees to take a stand and defend their beliefs by backing them up with data and information. It's OK to bend to management's vision, but you don't have to cave.

"Success as a manager or employee usually has less to do with your degree, your natural talent or even your intelligence," writes Bruce Clarke[36], president and CEO of Capital Associated Industries, Inc. "It has much more to do with your own personal level of persistence and determination."

In his writing, Clarke encourages people to keep speaking up and lobbying for what they want. Those who give up whenever they're told "no" will never move their agendas forward and will struggle to get noticed by the company.

Middle Managers Need to Create a Culture of Innovation

---

[35] Brendan Reid, www.brendanreid.com/blog-1/2016/2/2/my-best-tips-for-talking-to-senior-executives, My Best Tips for Talking to Senior Executives, Brendan Reid, February 2016

[36] LinkedIn, www.linkedin.com/in/bruce-clarke-50b975, August 2018

Before you can start pushing management out of the comfort zone, you have to know what you're getting them into. George Ambler[37], leadership writer, and executive in Nedbank's group technology division, illustrates those options with a target chart that breaks down the three places within a company that a leader can reside:

- The Comfort Zone: This is the status quo, where leaders spend most of their time.

- The Learning Zone: This is where leaders push the boundaries of their expertise to learn more and grow from their mistakes or successes.

- The Danger Zone: This is where the leader is stretched too thin and taking too many risks.

Using Ambler's model, you can better understand the importance of teamwork within a company. If everyone approaches management with new ideas and projects, it's easy to push them into the danger zone. By working together and prioritizing change, you can move management outside of their comfort zone without pushing too far.

John Marcante's Story

Often, it's up to middle management to encourage employees to innovate. Without leaders paving the way, entry-level employees will take Reid's "wet noodle" approach to pitching ideas, and upper management will stay in their comfort zones as much as they can.

---

[37] George Ambler, https://www.georgeambler.com/leadership-develops-when-you-escape-your-comfort-zone/, George Ambler, May 2015

John Marcante[38], CIO of Vanguard, says the age of the company and the industry don't correlate to innovation. He believes that change is 70 percent culture and 30 percent technology. By studying how Silicon Valley companies are nimbler and more innovative, he creates a culture of change within his company. Interestingly, the "scare tactic" approach seems to work as a motivator within Vanguard.

"We are never complacent," he says. "It is drilled in our heads to be paranoid." It's this fear of falling behind that keeps them constantly moving beyond their comfort zones.

Understanding Management's Motivation

One of the first steps in moving people out of their comfort zones is understanding why they're stuck there in the first place.

"Some people are meant to stay in their comfort zones," Angela Cochran[39], Associate Publisher and Journals Director at the American Society of Civil Engineers, writes. "They do great work and are extremely dependable. What they lack is a need to advance. For others, it's more of a paralyzing fear of change."

Interestingly, it might be easier to approach management that is paralyzed by fear than upper-level teams who are content with maintaining the status quo. This is because people who are content might be hostile to those who want to rock the boat, yet people who are afraid of failing will do whatever it takes not to. In the case of Vanguard, Marcante says the risks of staying put are greater than the

---

[38] CIO Magazine, https://www.cio.com/article/3119305/careers-staffing/being-pushed-out-of-your-comfort-zone-can-make-your-it-career.html?page=2, Brain P Watson, September 2016
[39] The Scholarly Kitchen, scholarlykitchen.sspnet.org/2016/05/26/a-shove-or-a-nudge-moving-people-outside-of-their-comfort-zone/, A Shove or a Nudge, Angela Cochran, May 2016

rewards of stepping out of your comfort zone as a motivator to move forward.

Implementing Idea Management Within Your Team

Once you know what motivates people, you can start to introduce ideas that speak to their problems.

Jeffrey Fermin[40] at The Vision Lab describes the concept of idea management and how a steady flow of ideas can create a culture of innovation and change. "When people are passionate about what they do, it will only spark more ideas and greater innovation," he writes. "It is up to your organization to make sure that people can give ideas without any hesitation."

This is where managing turns into leading. By fostering an environment of respect and listening, managers can encourage employees to contribute without fear of getting shot down — or worse, having their ideas co-opted and presented as someone else's.

Tailoring Your Requests to Management's Ears

Managers are swamped with questions, requests, and problems all day. Part of the reason you might not feel like you're heard is because it sounds as if you're just adding another problem to the pile.

"We always want to believe that our problems need immediate resolutions," Ryan Creamore[41], Strategic Sales at Trullico, writes.

---

[40] Vision Lab, thevisionlab.com/innovation-posts/idea-management-leadership/, Why Great Leaders Choose To Hear Everyone's Ideas, Jeffrey Fermin, February 2016

[41] LinkedIn, www.linkedin.com/pulse/5-tips-get-management-listen-you-ryan-creamore, Ryan Creamore, September 2018

"The first step in achieving your desired outcome is to remember not to bring problems; bring solutions."

When you present management with solutions, you're really asking their permission to move forward with the plan that you think is best. You're growing in your role as a manager, while the actual management team acts as a guiding hand, not like someone who is directly involved.

Guerric de Ternay[42], practical innovator and writer, agrees and encourages readers to understand what management wants and to frame their requests in a way that supports the company.

He presents remote working as an example. It's hard to sell your boss on the idea of remote work if you mostly want to skip your commute and wear pajamas all day. However, if your boss is worried about an early-morning call with a vendor in a different time zone, they might be open to the idea of your working remotely if it means you can take that call. Your goals are achieved by solving the problems of your bosses.

The FORTH Method

As you develop your requests and asks for management, it helps to create a plan for approaching them at the right time. Gijs van Wulfen[43], Linkedin innovation influencer, developed one strategy you can use that actually limits management's involvement to the beginning and the end of the process, called the FORTH method.

---

[42] Guerric, guerric.co.uk/convince-your-boss/ de Ternay, Guerric de Ternay, June 2018

[43] Idea to Value, www.ideatovalue.com/inno/nickskillicorn/2016/03/interview-gijs-van-wulfen-how-to-get-your-managers-to-listen-to-your-ideas/ , Gijs van Wulfen, March 2016

These five steps are ideal in getting management's initial buy-in then its full approval:

- Full Steam Ahead: Understanding the scope of a project and getting stakeholder buy-in.

- Observe and Learn: Developing a deep understanding of what the project needs.

- Raise Ideas: Brainstorming ideas to accomplish the challenge in an actionable list.

- Test Ideas: Evaluating what customers and employees think about your plan.

- Homecoming: Get internal support for resources and implementation.

Compare this process with the advice Creamore and de Ternay gave about coming up with solutions instead of problems. Upper management is typically involved in the beginning to get the go-ahead to move forward, and then again at the end when you have a finished plan for them to approve.

As a manager, this gives you the autonomy to develop solutions with your team and then present them in a way that gets approval.

Collaboration as a Persuasion Tool

Joel Garfinkle[44], the author of Getting Ahead, emphasizes the role of collaboration in getting support for your ideas.

"When you want to introduce a new strategy, schedule a series of one-on-ones with different people within the company — or even outside the company if that's appropriate," he writes. "Incorporate what input you can, and you'll stand a good chance of gaining their buy-in."

However, this is just the first step. The next step is throwing your support behind the initiatives of others.

By backing your co-workers in their initiatives and helping them when you can, you are more likely to win their support later on. This creates a strong web in which everyone builds each other up instead of treating management's attention like a pie that has to be divided up.

Putting Your Best Presentation Forward

Once you have the platform and confidence to present your ideas, you can focus on your presentation ability and tools. This requires a mixture of great data and a clear presentation to convey your message clearly and professionally.

---

[44] Career Achievement Blog, careeradvancementblog.com/ideas-heard-work/, Joel Garfinkle, November 2016

Tap into Imagery to Persuade

In 1986, university professors Richard E. Petty and John Cacioppo developed the Elaboration-Likelihood Model (ELM)[45] for persuasion. Essentially (not to be confused with the Kolb's Experiential Learning Model), the model suggests people learn through a pair of routes:

- A central route that handles facts and pros/cons of a decision;
- and a peripheral route that relies upon visuals to process a message.

Both play a factor in decision-making.

David Sands[46] at Adoreboard explains that this is where data visualization comes in. While management might listen centrally to what you're saying, peripheral visuals will help fortify your case.

"Part of the difficulty of introducing a new idea is ensuring that your audience understands just what it is you're trying to get across," writes Anett Grant[47], president and founder of Executive Speaking, Inc. "Given how much thought you've put into your idea, it can be frustrating when someone else isn't getting onboard and doesn't see the value or even understand what you mean."

She recommends using imagery — mock-ups, charts or photographs — to help people get on board with the overall vision. Even if

---

[45] Wikipedia, en.wikipedia.org/w/index.php?title=Elaboration_likelihood_model&oldid=853874740, Various Contributors, August 2018

[46] Adoreboard, adoreboard.com/visualisation-how-to-use-data-to-persuade/, David Sands, March 2015

[47] Fast Company, www.fastcompany.com/3045058/4-strategies-for-introducing-new-ideas-at-work, Anett Grant, April 2015

management isn't able to understand the pitch you're making, they can take in visual cues and infer what you want from them.

Highlight Incremental Value

"Show how your idea is better than the existing standard," Sarah Schmidt[48] writes at MarketResearch.com. This might mean pulling case studies from other companies who have gone before you or mapping out estimated incremental value.

Alternatively, you can use the scare tactic of mapping out what the company has to lose if management doesn't take action. How will it make your company lose money? How long will it take you to catch up as an innovator?

If your team doesn't respond to potential revenue opportunities, it is because they're comfortable with their current growth plan and highlighting potential lost revenue can help tap into their fear. Today's failure to innovate can be tomorrow's impediment to growth.

Be Brief, Be Clear

Maurice DeCastro[49], author of Hamster to Harmony: Get Off the Wheel and Live Your Best Life! emphasizes brevity and clarity when pitching to upper management.

Should you have an answer for all of their questions and have a thorough plan? Absolutely. However, you should present the big picture with your thought-out plans tucked to the side as a reference.

---

[48] Market Research, blog.marketresearch.com/how-to-sell-your-idea-to-senior-management, Sarah Schmidt, February 2016

[49] Hamster to Harmony, Paperbook, Maurice DeCastro, February 2009

Upper management is looking for a big-picture change and will typically only want more detail if they need to address a concern before moving forward.

At first, your management team will not like getting pushed out of their comfort zone. The risk to their careers and the company might be too high to agree to your ideas. However, with the right mixture of persuasion techniques and persistence, you can work to foster a culture of strategic risk and innovation to move the needle forward.

6 Ways to Make Organizational Change Easier

When you are in a meeting, and you hear the words "We need you to shake things up," push your chair back and run like hell. James Bond couldn't be more wrong with change management with his "Shaken – not stirred" approach. Organizations can go into outright rebellion to keep the old thinking in place. Additionally, if you hear "We are driving from New York to California. We don't know how we are going to get there but we will." What? In both of these scenarios, you are introducing chaos and a level of failure you will not recover from. You will hemorrhage talented employees, develop resistance to any new ideas, and everything becomes an outwit-outlast-outplay, no-win scenario. The only recovery from those approaches are to see when the CIO/CFO/COO leaves.

You need to set the stage to slowly introduce the new thinking into your organization with these six steps.

1. The Big Bang versus Small Steps

We've all had the pleasure – or displeasure – of being that person who led the charge to a better organization with eyes bright and chocked full of "Isn't change excellent!" all the while cluelessly going down a perilous path. Organizations don't like big changes or "Big Bangs." The larger the organization, the harder it is to control the change and communication out to the entire organization about the change (before and after it happens). Organizational change needs to be "stirred – not shaken." That means avoiding the big bang approach to changes and slowly rolling out small changes over a period of time.

Start with talking about one or two of the talking points above on failure. Don't overwhelm your audience with lots of ideas. Keep it simple and direct on how failing fast can help your organization significantly.

Once the concepts are clear, take on a simple process change that would help show your organization how failing faster is more effective. Typically, a retrospective that is well run is a good choice. Retrospectives allow you to take the problems that were discovered and prioritize those problems. Pick 1-3 problems and implement small changes to the organizational processes. Keep it small. Keep it focused. Don't try to boil the ocean.

Smaller changes are easier to handle and to digest for any organization large or small. The pushback on making small changes is "We have no choice but to make a large change." and the infamous "We need it all right now." Lies! We will call shenanigans on that. The size of the change is often linked to the implementation disruption size. The larger the change, the larger disturbance to the organization and the more likely the change will be viewed as negative. The trick is to take the large change and break it into smaller changes, so the organization is "stirred and not shaken." This takes careful planning and analysis on how to break things into smaller parts. Breaking into smaller chunks. Yes, this about to become a project.

Even with small changes to the organizational processes, you will need to communicate about the change regularly. Always talk about the value of the change first before introducing how it will impact the organizational processes.

2. Go Deep to Reveal the Impact of the Change

Go Deep. Understand how the organizational change impacts everyone. Validate the impacts it will have on every team. It might seem like your change won't impact some teams, but it's always a good idea to validate that. When organizations throw out change management, it is certain that the changes they want to make will not yield good learning. Let's take the example of making a change to a user interface or screen. The requirements are simple: add this field, move this field to the bottom and change the values in a drop-down box that user selects when completing the form. Simple right? Looking at the surface answer would be yes. You need to dig deeper.

That additional field will require training to the users on how it's used and why. Where is the data for the new field being stored? Will it appear on reporting? Include it in the data warehouse? How will the data entry be validated to ensure quality data is entered? What are the data validation rules for the new field? How does this impact the users of the interface when this field appears on their interface? Can we actually report on the data the way the business partners need?

The field that moved should be confirmed with users of that interface to ensure it will not break or place a burden on the existing data entry process. Users get in the habit of performing data entry without looking at times and changing up fields on the screen can cause some significant problems.

Amending the values of drop down also opens a lot of questions. What do the data values mean? What does the new data value mean? Can the existing database handle the new value? What about reporting and data warehousing? Will users need to be trained to understand the purpose of the new value and when to use it?

All the while you are addressing the impacts with the individual teams, you will be communicating the value of why the change is taking place. Never assume the individual or organization understands the value of the change. Ask them and ensure they are clear on the value of making the organizational changes. Moreover, if we go back to the beginning, did we understand the change they were asking for? Was it just a field they needed? Now take change management and massive changes to the organization x1000, and you have the real problem with overnight transformations and big bang theory changes.

3. Create Common Ground

One of the biggest reasons change fails in an organization is that the organization doesn't have a common ground in which to stand on. The objective is to create an environment or place where members of the organization can come to understand the change. This can be a website, but sometimes a physical location is needed. The personal contact of swinging by a room or office to learn more about an upcoming change is powerful in that it is far easier to resolve issues about an organization change in person than with email. A virtual meeting space that is manned over a period of several hours may be another approach.

Build a grassroots coalition with your colleagues but also reach higher in the organization. It's valuable to have a senior leader backing your change and seeing it of value for the organization.

4. Leading the Horse to Water

Organizational changes require a marketing plan. It's pretty close to the marketing plan your organization uses to acquire new customers. You will need to market the change to the organization. Your objective is to get everyone aware of the organizational process changes. Not everyone is going to be excited about the change as you are. Some may even be plotting to stop it all together. How would you use advertising to communicate with your organization? Remember you are in marketing. The idea is to get as many people in the organization (or the part of the organization) aware of the change. Advertising creates awareness very quickly.

Advertising is also about generating a consistently positive awareness of the change. Know your message and develop a 30-60 second memorized speech. Have it ready at all times to either cut and paste into an email or to recite it in the elevator on the way to the first floor to whoever is standing next to you when they ask, "What are you working on?" It's also important to figure out the user profiles. Based on those profiles (expert process guru, casual user, warehouse guy, or others) put together a WIIFM (What's In It For Me?) message. Think about how you could send that message more graphically or in an eye-catching way. Remember you're marketing, and you need these profiles to be aware of the change and to encourage them to be positive about it.

Blasting an email out doesn't always get attention from the user profiles that are impacted by the change. Email inboxes are always overflowing with new emails. Find a new way to communicate the message. One approach that Paul used on a team of 50 was walking around to each of their desks and dropping off a flyer with a little candy. Another way was to instant message a group and let them know of the change and that we would be available at a specific time for folks to drop by the desk with questions. This technique works even better when you schedule it multiple times throughout the day.

Beyond WIIFM factors, you have deeper profiles and analysis to be done as influencing without authority is always a factor. The more processes, people, and systems that need to change the more personalities and fiefdoms you will have to deal with. Contrary to what some books will tell you about influencing without authority, it does not solve everything. Most assuredly, it is not easy to do. Your marketing plan just got a lot more complicated. We are not saying this to scare you or dissuade you from pursuing change. You have to be in it to win it.

5. Culture is a Killer

If the organization's culture is averse to any change, then naturally the fight will be uphill all the way. As Peter Drucker said, "Culture eats strategy for breakfast." Continue to push the change forward and seek help from senior leadership to overcome roadblocks. In this scenario, you will spend almost all your time paying personal visits to key individuals to help you influence the importance and acceptance of the change within the organization. The "Snow Plow" approach of just pushing forward with a change won't work. Changing an organization's culture takes a long time. You will not be able to force it overnight. It could take years before the change is fully deployed. Culture change is dependent on the size of the organization, the commitment from top leaders, and the resources available to make it happen. When a change is forced and the resources are not available, the change will not happen fully. People will eventually go back to their old ways. Studies (Barry Phegan, Ph.D, 2018)[50] show that just 10

---

[50] Company Culture, companyculture.com, What is Company or Organization Culture?, Barry Phegan, Ph.D., August 2018

employees could take a full year to change. 100 takes 2 years. 1,000 takes 3 years. 10,000 takes 4 years and so on.

It is no wonder why organizations flip-flop so quickly on Agile and other approaches. They don't have the resources or the top leader support. The size of their organization cannot match the expectation of management for change. I.e., Management expects change to happen in a year in an organization with 10,000 or more employees.

6. Fast and Furious is a Movie – Not an Organization

Large organizations by their nature don't move quickly when it comes to change. A team of 5 can absorb change far quicker than a group of 50. The more significant the impact to the organization the more time you will need to plan the change and communicate it. Keep this in mind when setting expectations. Change takes a lot of work.

Building Team Resiliency to Navigate Change

Teams can build resiliency by working closely with managers and executives to navigate change smoothly. As your organization gets used to change it learns to trust you through the process, and it becomes easier for the organization to move forward.

Navigating change requires soft skills like communication and empathy, along with high EQ (emotional intelligence) levels. Guess what Agile and DevOps embrace? Empathy and EQ. Guess what gets chucked on most Agile and DevOps teams when they start? Techniques and approaches that promote empathy (empathy maps, personas).

Why do employees fear change? A fear of change isn't limited to the workplace. Most people try to prevent change and worry about changes they can't avoid whether in their personal or professional lives. The fear can manifest itself differently in an office environment and can affect those around you.

Employees have different "Psychological Capital" levels. As you lead your team through change, you are likely working with employees with varying "psychological capital" or "PsyCap" levels. This will determine how your employees face different challenges.

In her article at The Conversation, Yvonne Brunetto[51], who teaches management and HR at Southern Cross University in Queensland, explains what makes up someone's psychological capital:

- Self-efficacy: how confident or self-assured someone is.

- Optimism: how positive someone feels about the future.

- Hope: how determined someone is to work hard toward a goal.

- Resilience: how effectively someone can bounce back from a difficult event.

These factors work together, creating some level of PsyCap for everyone. Moreover, each factor is important: Even highly-resilient employees can be a drain on the team if they have no hope or optimism for where the company is going.

---

[51] The Conversation, theconversation.com/can-you-teach-workers-to-be-more-emotionally-resilient-63190, Yvonne Brunetto, August 2016

You can build optimism, hope, and resilience by creating clear concise communication that shows the value of the change and working with your organization at all levels to build trust and support for the change.

Employees have multiple reasons to avoid change. Along with varying PsyCap reactions, employees also have different reasons to worry about impending changes. The concerns they experience depend on their backgrounds and emotional insecurities.

Robert Tanner[52], business consultant, and development coach, highlights a few common reasons why employees fight against change or stress out when something new is on the horizon:

- Loss of status or job security. The change may affect their place in the organization.

- Fear of the unknown. The less your team knows about the change, the more fearful they will become, filling in the blanks themselves with rumors and assumptions.

- Organizational politics. Some people resist change to prove that it is wrong for the company and try to lead others to do the same.

Tanner says people don't resist change when they believe it's in their best interest. When presented carefully, teams can embrace change as a positive step for the office rather than fight against it or try to avoid

---

[52] Management is a Journey, managementisajourney.com/organizational-change-8-reasons-why-people-resist-change/, Robert Tanner, August 2018

it. There is something to be said for politics and the big bang theory of change it represents and the resistance it gets.

When trying to get an organization to embrace failure, it's important to show the value in why failure works for an individual and organization. The Experiential Learning Model is a proven model for learning and demonstrates the value of failure to the organization.

"[Some employees] want to feel needed and indispensable to the department or organization," Nancy Anderson[53], job search and career coach, says. "If they are the expert with a particular system, they feel pushed out when a new system is introduced, and the old one is scrapped – it puts everyone on the same level, and they become a less vital member of the team for a while."

Expect initial resistance to your changes. While executives expect stress and pushback from major changes, like team adjustments or downsizing, most leaders aren't prepared to handle change on a smaller scale.

Small changes that demonstrate the value for embracing failure is an excellent way to address roadblocks, but there are some roadblocks you might not be able to overcome easily even with excellent communication and a clear value statement. Expect that there will be some in your organization that will always resist the organizational process change no matter what.

---

[53] People®, www.peoplehr.com/blog/index.php/2017/06/07/why-employees-resist-change-even-when-its-good-for-them/, Why employees resist change – even when it's good for them, Nancy Anderson, June 2017

Create a response plan for resistance. Employee resistance to change can have immediate impacts. Rick Maurer[54], the author of Beyond the Wall of Resistance, reports that 80 percent of CIOs say resistance to change was the main culprit of failures in the organization, even beating out a lack of skills. Most Fortune 500 CEOs agree, citing resistance as a significant reason that changes they tried to incorporate failed.

"How you respond to employee complaints and initial pushback to change can set the tone for the entire transformation," change communications consultant Beth Archie[55] writes. "If a few complaints come to HR or someone sends an angry email to the CEO, and it brings change to a halt, you could end up creating a machine that makes your employees better and better at resisting future change."

Leaders who stop change based on some initial resistance are essentially rewarding employees for pushing back and proving that all objections are immediately valid and will be acted on.

Vicky Webster and Martin Webster[56] at Leadership Thoughts suggest engaging with employees who are resistant to change and discussing how the change will affect them and how those concerns can be alleviated.

---

[54] Rick Maurer, rickmaurer.com/wrm/, Resistance to Change – Why it Matters and What to Do About It, Rick Maurer, September 2018
[55] Spark, www.adp.com/spark.aspx, Do Your Employees Resist Change?, Beth Archie, August 2018
[56] Leadership Thoughts, www.leadershipthoughts.com/how-to-recognise-the-signs-of-cognitive-resistance-to-change/, How to Recognise the Signs of Cognitive Resistance to Change, Vicky Webster and Martin Webster, August 2018

In particular, they encourage people to commit to at least a small contribution to the change. Taking this one step can reduce fears of change as employees don't have to make a big change all at once. It also turns an adversary into an ally as they take small steps in favor of your plan. Breaking the change into smaller pieces helps you prepare for this ahead of time, so you are not facing a roadblock unexpectedly.

Execute organizational process changes effectively. When you announce an organizational change, half of your team is likely to feel excited and optimistic about the opportunity, while the other half is convinced that their careers and company futures are in jeopardy. Knowing this, you can create a successful plan for change that benefits the majority of your staff

Most people follow the FIRE method of information processing, whether they realize it or not. Mark Murphy[57], the founder of Leadership IQ, says understanding how people use the four-step FIRE way of thinking can provide insight into how your employees process change.

- Employees are given some Facts or information.

- They Interpret those facts for context.

- Those interpretations are often built on emotional Reactions.

---

[57] Forbes, forbes.com/sites/markmurphy/2016/08/14/the-big-reason-why-some-people-are-terrified-of-change-while-others-love-it/#6a84864c2f63, The Big Reason Why Some People Are Terrified Of Change, Mark Murphy, August 2016

- Those emotions lead people to determine the expected or desired End.

FIRE reactions often occur when the change has high levels of uncertainty. For example, team members might hear that a new tool will save time and require fewer resources. "Fewer resources" will be interpreted to mean fewer employees will be needed. This is a fearful emotional reaction that leads people to resist using the new tools or pushing back against any changes.

Carefully present the change news to your organization. To prevent FIRE reactions from blazing out of control, leaders can strategically plan how they want to present and roll out changes within an organization.

Megan Maslanka[58] at Quantum Workplace encourages both team leaders and employees to focus on the facts. Those who significantly fear change often imagine worst-case scenarios in unfamiliar situations. Management can reduce water cooler talk by clearly presenting what is going to change and how those changes are expected to affect people.

That being said, you want to avoid presenting too many details at once. Workplace motivation expert Jeff Miller, Ph.D[59] shares his own

---

[58] Quantum Workplace, www.quantumworkplace.com/future-of-work/3-tips-to-overcome-fear-of-change-in-the-workplace, 3 Tips to Overcome Fear of Change in the Workplace, Megan Maslanka, September 2017

[59] Inc, www.inc.com/jeff-miller/how-to-lead-employees-through-times-of-change-at-work.html, How to Lead Employees Through Times of Change at Work, Jeff Miller, Ph.D, September 2018

story of getting laid off. After hearing the news, his head started spinning, he couldn't focus, and he was so overwhelmed by the news that he wasn't able to digest the information following it. The emotional aspects of the news made it so he completely forgot questions related to severance pay or recommendations. The same can be said for employees. Changes can be so stunning that they need time for the news to sink in before they can process the details.

Employers that try to sell a change idea to employees or inundate them with details could overwhelm their team. You should certainly have a plan to enact the change, but also give your team time to process what is going to happen.

Foster a resilient company culture. Change is inevitable, and you can navigate change successfully by building a resilient team. Resilient teams still need guidance through change, but they are often less change-phobic. There are a few ways you can encourage employees to help themselves become more resilient to change. Here are some good practices to follow:

- Develop healthy friendships and relationships.

- Keep a level head and maintain the perspective of the situation.

- Nurture a positive self-perception. See yourself and the change as positive.

- Take time to recharge and step back from the situation. Don't push yourself into exhaustion.

Additionally, you can foster resilience in teams during change by becoming a resource to your colleagues. Leaders build organizational resiliency by empathizing with their teams and helping them through difficult challenges. Be there to help them through the organization process change.

Being a Change Agent for Failure

"The measure of intelligence is the ability to change."
- Albert Einstein

You need an army of change agents. A few won't do in a large organization. Before you say we don't need change agents because it is just about communication, keep in mind that most people are terrible communicators. A change agent for an organization doesn't have to sit on the board of directors or be a senior leader. Anyone in the organization can create change. Not everyone is successful with being a change agent. Here are a few characteristics that every good change agent has:

They have a clear vision of the change, and how it affects the organization. There is a compelling story for why the organizational change is of value and how it will help the organization. Know your story well. Have a clear and concise message memorized that explains the change and its value.

Change agents are patient and persistent. Organizational change takes time, and no organization changes overnight. Remember the statistics on how long it takes? Get used to this idea right now. Change agency is something everyone needs to know in your organization and it is never going away. Pony up the cost to deal with it right now. You may only be able to move the organization a small step towards the overall vision of accepting failure and harnessing it in your organization. Don't be discouraged. If you can move the needle towards your vision – even in a small increment – that is a huge success. Don't waver in your persistence and don't assume everyone is on board with the vision. Keep communicating about the value of the change.

Question everything and dig deep. Find the deeper meaning of the change and the effect it will have on the organization. Understand how harnessing failure would work in your organizational processes profoundly and thoroughly. Demonstrate your understanding to others to build trust that you are looking out for their best interests and understand how the vision will change the organizational processes.

Change agents lead by example by showing how they adapted to new changes in a positive way. They support their colleagues with the change and help them adjust to the change. Be pleasantly helpful when others run into trouble with the change. Ensure the value of the change is understood, so your colleagues don't give up on you.

Change agents build trust and good relationships with all levels of the organization. Making a good connection to all levels in the organization is critical to being successful with organizational process changes. These connections significantly assist in getting the organization to adopt new attitudes towards failing fast and learning from failure. The Pitfall Principle is a significant player in building trust. This principle states that others will respect and trust you if you admit your own mistakes and learn from those mistakes. When you fail and recover from failure, you are demonstrating the principles of fail fast in real life. When other see those fail fast principles in action and how you learn from them, they are more likely to embrace those principles.

Support Others

As you sought others to support your experiments and failures, you will also need to encourage others in your organization that have proposals for their experiments. Support others in your organization where you can to foster collaboration within your organization on experiments. Failing safely needs to have a network of supporters even after the boss has approved the experiments.

Support of others in the organization with their experiments creates new opportunities for collaboration at many points along the fail fast model.

In discovery validation phase of the Fail Fast model, having colleagues or others reviewing your understanding of the discovery gives you a broader perspective of potential experiments to design. Additionally, this broader perspective gives you a better viewpoint to determine if the discovery is worth pursuing.

In design experiments phase of the Fail Fast model, having colleagues assist you in reviewing or designing your experiments ensures those experiments are formulated to address critical issues about the discovery. Input from others helps you focus your experiments. Additionally, input from others might see different techniques and methods that could be used to address the discovery.

In conceptualize improvements phase of the Fail Fast model, observation of the concrete experience by a broad range of viewpoints strengthens the ability to find new innovative approaches to resolving the failure that occurred in experiments.

Having others involved in Viability Analysis phase of the Fail Fast model involved in the discussion about whether or not to continue experimenting is also essential to helping you know if your experiments should cease or if new approaches should be embraced.

There are even more ways that have a support network can help you safely fail that are not listed here. Build a strong network of individuals in your organization to create a better environment for Failing faster and safer.

Create Awareness - Address Experimentation and Why It's Good

Communicating how failure works and the reasons it works is an important step of awareness. Others in your organization will be more interested in and pursue experiments when they understand why the fail fast approach works to foster learning and innovation.

Educating others on the science behind failing and experiential learning will help others in your organization understand the benefits of experimenting and failing fast. Experimenting isn't just playtime for the sake of playing. There are real benefits behind experimenting to improve products and services. The time spent experimenting is worth the potential innovations that could occur. Experimenting helps provide individuals with a deeper understanding and insight into tools, products, services, or whatever.

Active experimentation doesn't typically occur in the customer-facing environments but rather behind the scenes. Address this fear up front in that experiments are run privately and not publicly in from of customers.

Here are some example talking points:

- Failure isn't dangerous – it helps us learn more quickly and deeply
- Experiments play a crucial role in framing out learning
- Critical Reflection on the failure helps determine the root cause for future improvements
- Concrete experiences with the product ensure a deeper understanding of the product
- Experimenting helps create new potential opportunities for product improvement
- Experimenting builds product knowledge
- Experimenting creates a creative mindset and makes innovation faster

Would you add other points to this list? What points would you not talk about and why? Every organization, team or group of individuals are going to be different. Craft your message to address the specific desires, fears, and issues.

Set Expectations

Once you get the "go for it" approval – you will need to make sure to set expectations around the experiments – what can fail, reporting back results, how you will fail safe, and valuable lessons learned are communicated out. Communication is key. Sponsors, executives, management, stakeholders, and others in your organization need to feel comfortable with the experiment and the potential outcomes. They need to fail safe.

Setting expectations will allow the organization to fail safely. Being upfront about the experiment, it's purpose, and what happens when it fails are critical parts to ensure there is a feeling of safety about your experiments in the organization. Don't hold back bad news. Provide transparent information no matter the outcome of the experiment.

Here are some examples of thing to communicate on the experiments and failure:

- We know we will fail, and we are trying to learn from failing.

- We are failing safely by taking steps to ensure the failure is not public and done in an environment that is conducive to experimentation.

- Show the benefit of failing fast - think Silicon Valley – think about Agile and being more flexible.

- Failing fast provides us the opportunity to develop innovative ideas quickly.

- Hands-on or concrete experience gives us the best understanding of our product.

- Allay fears where you can. Overwhelming fear stops creative and innovative solutions dead in their tracks.

- Make sure you have plan B and a soft place to land.

Chapter 5

Fail Fast Model
Part 1 - Initiate

"I like failure because it's so easy to achieve!"

- Anonymous

In this section of the book, we take the theory behind experiential learning and turn it into the Fail Fast Model. Let's walk through the Fail Fast Model to understand each of the model's steps. We will then apply the model to a real-world example.

The Fail Safe Model

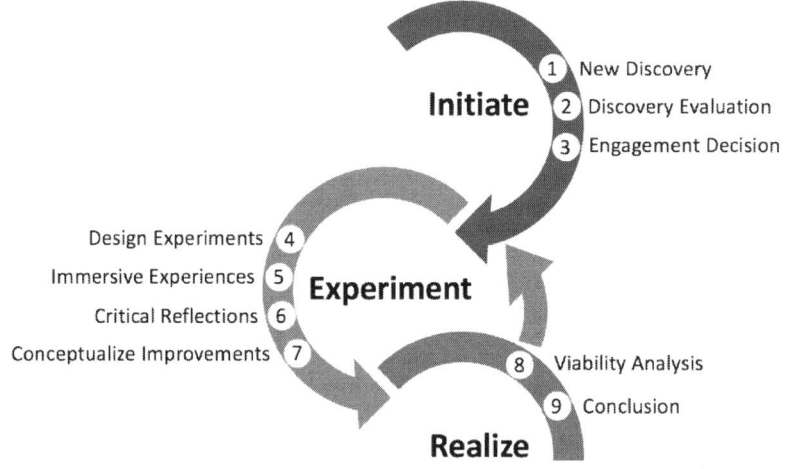

Some questions you might have as you go through the model:

Isn't this just an R&D approach for products?

No. Research and Development teams use a more robust approach when designing new products or making changes to existing products. In the world of business and technology where products are more service based, this model can be used and holds merit. As an example, the establishment of success criteria for a service-based product is equally vital to setting success criteria for a physical product like a toaster.

Is this model just a Quality Assurance model?

No. There is more to failing fast and failing safe than just testing. However, experimenting and testing a product or service against success criteria are a vital part of the model. Also, designing experimentation and conceptualizing improvements don't fall within the typical Quality Assurance focus. Immersive experiences and critical reflections do overlap the quality assurance models slightly.

Is this model a methodology?

We designed the model as a way of thinking and learning – not a specific methodology. A methodology is defined as a set of practices, procedures, and rules that are used by a specific group of individuals or a particular career discipline. We designed this model to be more open to a larger group of varying disciplines as a way of understanding how failing fast helps you to learn faster. If you look a little deeper into the model, you can see it is a rationale of why some methodologies work better than others. We always subscribe to doing what is best for you. Take the pieces that make sense to you. The others will be there for you when you are ready.

The Fail Fast Model and Change

The model is meant to change. The model itself can fail, and it should fail. We urge you to consider making changes to the model and experimenting with those changes to get better results for you and your organization. It's okay to rethink the model and change it. So, if changing the model will help your organization fail faster and learn more, we are all for it. We don't want you to feel it's our way or the highway. There isn't any single model that will work 100% of the time in any organization.

Initiation Phase

Initiation Phase is where it all begins. It is where a new situation or opportunity materializes. The Initiation phase contains three parts: New Discovery, Discovery Evaluation and Engagement Decision.

1. New Discovery

Discovery starts when a unique situation or opportunity is:

- Observed by noticing something unique or new. You are walking along and see something you have not seen before.

- Presented to you by a friend or colleague and is requesting your opinion or assistance. A friend is seeking help in solving a problem. Your friend or colleague asks for your advice or point of view.

- Presented to you by someone in authority who is directing you to engage with a situation. Your employer or manager is requesting you to perform a task, activity, a project, or business case creation.

- Self-initiated by trying to fill a knowledge gap during research. You are presented with a discovery while you are performing research on another task, activity, or project.

- Trying to resolve or find a solution to a problem. While working to solve a problem, you discover there are more facets or parts to the problem.

Many models will help with New Discovery. We will look at two specific scenarios to demonstrate some of the models that can be used:

1. Corporate environments - looking for opportunity.
2. Self-discovery – looking for personal improvements.

Models for New Discovery Corporate Environments

Here is a list of models that can be used to drive out new discoveries:

- Environmental Scan
- Business Model Canvas
- Agile Vision Board
- SWOT Analysis
- Brainstorming
- Focus Groups
- Porters 5 Force Analysis
- Workshops
- Organizational Modeling

- Current State Analysis
- Future State Analysis
- Context Diagram
- Functional Decomposition
- Financial Analysis
- Benchmarking
- Market Analysis
- Vendor Assessments
- Metrics Analysis

- Interviewing
- Business Capabilities
- Data Mining
- Risk Analysis
- Surveys/Questionnaires
- Lessons Learned

- Item Tracking
- Concept Modeling
- Process Modeling
- Root Cause Analysis
- Observation

Let's look at The Environment Scan for discovery in a corporate environment.

The Environmental Scan

The environmental scan is typically used in Enterprise Analysis or what is known as Strategic Analysis, to determine what internal or external factors might be impacting your organization. The basic purpose is to discover and help manage the future direction of your organization. It is a strategic approach to discovery that goes beyond a typical SWOT (strengths, weaknesses, opportunities, and threats) analysis. It can capture trends in your organization, identifies demographics, market impacts, tech needs/barriers and more depending on how you use it. It is a surprisingly flexible technique and can be used to find gaps in your business practices, systems or even cancel projects (Bob used this technique to cancel a 23-million-dollar project).

When Should I Use It? Why Would I Use It?

Use it in Enterprise Analysis, when developing ideas for business cases, searching for improvements, and to search for redundancies. Also used as a self-check mid-project, when the project seems outmoded, old, or you don't think it will work, and you need firepower to back up your arguments. There are tectonic shifts that happen in your organization due to the rapidly changing business and technological advances that impact your people, process, and systems. Tectonic means we relate to the structure of the organization, and the large-scale processes that take place within it.

Preparation

Items you will need to assist you in this technique:

- Whiteboard or flip chart
- Markers or pens
- Sticky notes
- Spreadsheet on your computer
- Identify your time box for the effort. Shorter for less discovery and longer for more.
- Based on the topics at hand, identify the appropriate individuals to collaborate with on this technique. You will likely need a diverse team; roles, backgrounds, skill sets, expertise.

How to Get it Done

Let's start with what this technique looks like:

FACTORS	Going Away	Generally Accepted	Emerging	On the Brink
1. Technology				
2. Culture				
3. Politics				
4. Demographics				
5. Economics				
6. Environment				
7. ???				

Yes, it looks like a spreadsheet, but it is a spreadsheet with powerful discovery waiting to be found! It can be done on flipcharts and whiteboards too.

4. First, you must consider the factors (down the left side of the spreadsheet), Typical industry factors include:

    a. **Technology** - Systems, tools, concepts.

    b. **Culture** - Typically meaning the culture of an organization – internal culture. Culture could also be used in terms of a market, region, state, country or the world.

    c. **Politics** - Internal organization politics but may include 3rd party vendors, merging companies or acquisitions.

d. **Demographics** - Internal organization demographics. Roles, gender, career path.

   e. **Economics** - Internal economic factors. Salaries, 401K, benefits, retirement plans, HSAs.

   f. **Environment** - Typically viewed as soil, water, climate, and vegetation. In an organization, this is likely to be workspace, tools, and other physical needs.

   g. **???** - Additional factors can include social, legal or other organizationally driven elements that are important to capture.

5. Next, you must understand the categories that your scan will be balanced against.

   a. **Going Away** - Going away means that it is going away in your organization (or context of your scan). It is no longer being used or going to be used, being offered, or it is thought of as an antique or outmoded.

   b. **Generally Accepted** - Generally accepted means that you agree with the idea, concept or statement being made. It does not mean you have to like it. The organization agrees it is the norm.

   c. **Emerging** - Emerging are things that some people are using but the broader population (again – in the context of what you are scanning) is/are not using.

d. **On the Brink** - On the brink are things that people are talking about, but no one can use yet. It is the future of what could be. It is coming but when?

6. Let's walk through an example. We will scan the world. What is going away in the world with technology? What comes to mind? Let's identify technology that is no longer being used. The list could be quite long. CDs? Landlines? Flip phones? Fax machines? All great examples, however, now let's go a little deeper. If you are working for a government agency are fax machines really going away? Probably not. So, this means that this scan has to consider the organization and people that you are working with because 'going away' for one company could be 'generally accepted' at another. If fax machines are no longer in use or will soon be, your organization would categorize as going away. If you are still using them and a sunset date has not been set, they are generally accepted.

What is generally accepted about technology? Well, it may be fax machines. What else? It is always changing! Wait a moment. Notice how we just went from physical objects in going away to a concept in generally accepted? The scan covers more than one type of result. Physical or conceptual. What else is generally accepted? Systems are getting more complicated to work in. Systems get smaller like cell phones. We can't keep up – it is out of date as soon as we buy or install it. Fax machines!

What is emerging in technology? 3D printers, driverless cars, AI. Those are all examples of things some people are using in the world but not many people. 3D printing has become very accessible. You can buy 3D printers at lots of stores. I don't know many people that have purchased 3D printers so this is still emerging and not generally accepted.

What is on the brink with technology? Space travel for everyday people, holodecks, humans with embedded knowledge chips. Scary and shiny and all things people are talking about, but no one has access to or are using yet.

FACTORS	Going Away	Generally Accepted	Emerging	On the Brink
1. Technology	CDs, land lines, flip phones.	It is always changing, systems get smaller, already out of date, fax machines.	3D printers, driverless cards, AI	Space travel for everyday people, holodecks, humans with embedded chips.
2. Culture				
3. Politics				
4. Demographics				
5. Economics				
6. Environment				
7. ???				

7. Now lather, rinse and repeat through the other categories.

Once you have it filled out, you must use this to drive changes in your organization. Here are two examples of how this scan could have been used and was used.

Borders Bookstore. They did not do an environmental scan. They invested heavily in CDs and did not embrace the idea of digital downloads. Borders Books does not exist anymore. An environmental scan might have prevented or altered the outcome of their demise. If they had scanned the technology industry, their organization, and their competitors, they would have seen it coming. They did not see it coming, and we got some great bargains when they closed.

Bob was on a project that did not "smell" right, so Bob did an environmental scan. It was determined a significant amount of money was being invested in technology that was not only going away, but the business partners were not going to be able to support. It still took some time to convince people, but we used the environmental scan as a business case to get the project canceled. We used it to explain sunk cost fallacy and opportunity investment. It was the right thing to do. It saved the company 23 million dollars that could be invested elsewhere and on the right things.

Our Challenge to You #1 - Finish scanning the world.

Scan the world based on the other categories. Additionally, create one more category (to replace the ???) and scan on that too. It can be anything that makes sense. Something that is not already been represented in the factors.

Our Challenge to You #2 - Scan on Your Organization.

See what is happening in your organization and industry. You don't want to become Borders Books. You might find some surprises. You might just validate what you know. In either case, you can use it to help your organization fail faster with this discovery.

Our Challenge to You #3 – Scan a Project, Process, or System

Scan a project, a process, or even yourself. That is right; you can scan at many different levels including things, products, people, or just about anything. Just change your lens a bit. Go more narrow. What processes are going away? Why? What processes are generally accepted? Why? Should they be enhanced? What processes are emerging? On the brink? What do all of these things mean?

Final Thoughts on the Environmental Scan

Some advice on using this technique. Not all categories are that easy to fill in the information on. It is dependent on subject matter expertise. You may not be able to fill out ON THE BRINK because you have no knowledge of things to come. Seek out an architect or another visionary in your organization. You can do this technique solo or in groups. It might make more sense to do it in a workshop. If your results yield nothing, it means you are doing great! Be prepared to find many things though. The real work begins soon.

Models for New Discovery – Looking for Personal Improvements

Many of the models listed in the discovery section could be used to find personal improvements. Doing an environmental scan on oneself can be pretty powerful. You will get a very good idea of what is going on in your life. Before you try out the environment scan, let's take a closer look at the SWOT Analysis. It will be a simpler approach to start with when analyzing yourself.

SWOT Analysis

SWOT Analysis is typically used as a strategic planning technique in an organization to identify the Strengths, Weaknesses, Opportunities, and Threats related to business competition. What if we used it differently? Let's use it on ourselves instead of an organization. Let's use this technique as a tool for introspection and personal improvements.

When Should I Use It? Why Would I Use It?

Whenever you are stuck and not sure of your career path, your growth opportunities, or what might be getting in your way of achieving success.

Preparation

- Whiteboard of flip chart
- Markers or pens
- Sticky notes
- Spreadsheet on your computer
- Identify your time box for the effort. Shorter for less discovery and longer for more discovery.
- Identify key resources that can corroborate your findings from this technique.

How to Get It Done

Let's start with what this technique looks like:

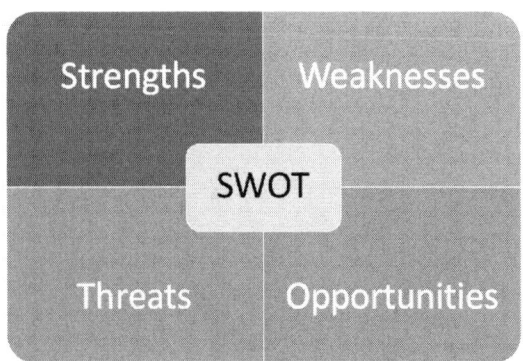

A simple grid structure that breaks down the acronym of SWOT: Strengths, Weaknesses, Opportunities, and Threats. Strengths and Weaknesses are internal to you. Opportunities and Threats are external to you. This model is intended to produce a simple one-page diagram.

How to Complete the SWOT Model

Step 1 – Copy the SWOT template above to a piece of paper or copy it to a whiteboard.

Step 2 – Write down your strengths. Brainstorm your ideas and write them down. What makes you great? What are the things that you do well? What are the things that drive you? If you have not taken Strengths Finder 2.0 by Tom Rath, that would be a significant additional step to take. It has made quite the impact on Bob's life. Are you a great communicator? Are you the harmonizer of the group? Do you influence well? Strategy? Details? Understanding our strengths is essential, so they can be balanced against our weaknesses. Example: Bob's biggest strengths are winning over others (Woo) and strategy.

Step 3 – Write down all of your weaknesses. What is it that you don't do a great job at? Where do you need improvement? What do you wish you could do better? Example: For many, many years Bob put "communication" as both his greatest strength and greatest weakness. Talk about strategic! A calculated approach to put himself in more difficult spaces to practice his craft. Think about it. If Bob were less strategic and were more practical about what he wanted, he would put things like "easily gets too upset and takes things too personally" or "sucks at data modeling." Instead, Bob used the strategy to allow himself more opportunity for failure and experience.

Step 4 – Write down your opportunities. Knowing what you are good at and what you are not good at will naturally create an opportunity list. Example: This is where Bob's strategy kicked in. Know where your weaknesses will lead you. Rather than put weaknesses that would send Bob to a data modeling class or an emotional IQ class, he went to advanced facilitation classes and got to facilitate workshops and other opportunities that he did not have before. Now, you don't have to be that strategic with this model. Just use it for what it is.

Step 5 – Write down the threats to your success. What threatens your success? What will get in the way of your goals? People? Budget? Time? Process? Compliance? The mentality of "we have always done it that way" attitude? Example: Bob needed to make the case why it was essential to invest in his personal growth. He had to consider tradeoffs. He would have to teach others. He would have to prove out the ROI of taking a class. Bob would have to consider other individuals who want something as much as he does. How could Bob make them (the people that want things too) a part of the process and create a win-win for multiple people?

Step 6 - What can you see about yourself that you have not seen before? What do you see that validates what you knew already about yourself? Does seeing it give you the push over the edge to move forward on it? Example: In Bob's example it confirmed what he knew already about himself, but he also used what he knew about himself to uncover new things. Sometimes the things that we discover about ourselves are not expected or wanted but do move us forward. One could argue that a SWOT analysis is an excellent tool for all relationships.

Our Challenge to You #4 – SWOT Yourself

It's time to SWOT yourself! That is SWOT and not *swat*. See what you can discover. Where will it take you?

2. Discovery Evaluation

When you discover something for the first time, your mind tries to see if the discovery is something you previously encountered or familiar to you. If the discovery is familiar or previously encountered, you can make a quick evaluation of the discovery to determine if it is a threat or non-threating. You are also able to determine whether or not you are interested in the discovery. Because the discovery is something you previously encountered, there is a good chance you will ignore it without further thought. The human mind is designed to look for discoveries and evaluate those discoveries continually. This continuous searching happening in your brain is why curiosity is an integral part of discovery. Curiosity encourages you to not only seek discoveries but to evaluate everything around you. We dedicated an entire chapter to curiosity later in this book.

Evaluation is something you perform at almost a sub-conscious level, and it happens in milliseconds. Evaluation is a high-level overview of the discovery based on your previous experiences. The more experiences you have, the more able you are to evaluate a broader range of discoveries quickly. Think of the first time you did something like taking an elevator. Since you had no experience with the elevator, you were very unsure about how to interact with it. All of your senses are on high alert at this point. Your brain performs a threat assessment of the elevator by trying to match the elevator to previous experiences and discoveries that were similar to the current situation. You observe how others around you interact with the elevator and notice that others are not afraid of it or running away. Your senses are still heightened, and you are nervous.

When you get in the elevator, you assess the purpose of the buttons on the wall and determine that pressing one of those buttons will take you to the floor you desire. You noticed others pushing the buttons and act in the same manner. As the elevator doors close and the elevator starts to move, your heightened senses are registering everything from the pressure on your feet to the noises you hear. All of those sensory inputs are filed away in your memories for later retrieval. As the experience progresses, your mind is cataloging everything around you for the next time you encounter an elevator.

Based upon everything that happened with the elevator interaction, you have determined the elevator is reasonably safe and get more comfortable with it. You evaluate that the elevator is useful and non-threatening to you. Unless you have had different experiences working downtown in the business district in a high-rise, and you got stuck between floors 35 and 36 with the door open. Your evaluation will be entirely different.

Discovery evaluation can occur without you even thinking about it. It happens in milliseconds. Most people aren't aware the evaluation has occurred.

The experience is different when you are familiar with a discovery. Your senses are not as heightened. Your mind is thinking, "Been there – done that!" Previous experiences with a similar discovery come to mind. You begin to apply those previous experiences to the current situation. You perform Evaluation based on good and bad experiences. "Oh no not again" or "I love this stuff" quickly come to the forefront of your thoughts.

In the business world, we typically label Discovery Evaluations as high-level analysis or feasibility analysis. If the discovery is something no one has ever experienced in the organization, the same threat assessment that an individual performs now happens in the minds of individuals within the organization. An individual has a wide range of experiences that are both good and bad that can be brought into the discussion. During discovery evaluation, the organization is trying to determine more about the discovery. Organizations (like individuals) have the desire to get comfortable with the discovery. The positive and negative experiences of the individuals in the organization are pooled together. Ensuring an individual contributes their experiences about a discovery can help organizations better understand the discovery at a high-level more quickly.

Example: you are the Product Owner for a product. A problem has been discovered with the product when a customer makes a complaint that specific functionality isn't working. As the Product Owner, you evaluate the complaint and perform some high-level investigation that tries to repeat the customer experience.

Once an evaluation is completed, you move on to the engagement decision. The primary outcomes of the Discovery Evaluation phase are:

- **Problem or opportunity statement is created.** The problem statement defines what the problem is and its effects on the organization. The opportunity statement provides the rationale of why the organization is considering the experiments that will add value to the organization such as increased sales, profitability, or market leadership. Problem statements are reactive, and opportunity statements are proactive. Negative and positive are another way of looking at them as well.

- **Establishing high-level goals and objectives.** The goals and objectives that provide a focus for creating success criteria for your experiments. Traditional approaches to goals and objectives include SMART which are the criterion for judging your goals and objectives. SMART stands for Specific, Measurable, Achievable, Relevant and Time-bound. We like SMARTER which includes E=Ethical and R=Recorder. You can always go deeper and use John Whitmore's SMART-PURE-CLEAR approach.

  Providing enough information to decide. Provide enough background, value statements, and high-level information so an informed decision can be made to move forward.

Models for Discovery Evaluation – Corporate Environments

There are lots of ways to document goals, objectives, your evaluation, and justification. Our favorite is the Business Model Canvas.

Business Model Canvas

The Business Model Canvas was introduced by Alexander Osterwalder and has quickly caught on due to its simplicity, ease of execution and the ability to "canvas" your needs. It is a one-page view of what you want to accomplish and how you will get it done at a very high level. It is a mini-business case. Why do we like this model so much? It keeps it simple, and that will help you to fail fast and safe!

The Business Model Canvas is a template that provides insights you will capture about your effort including:

- **Value Proposition** - What is it you are trying to accomplish? What value are you trying to deliver to your customer?

- **Key Partners** - Who are the people that we are trying to please? Who are the people that are important to work with? Partners are not always customers or resources.

- **Key Activities** - What activities are accomplished with our value proposition?

- **Key Resources** - What resources are needed to make this effort work? Resource are people, systems, and tools. The people needed to get the work done may be different from your key partners.

- **Customer Relationships** - Who are the customers that we are serving? Customers are not always partners or resources. How do we grow the relationships with our customers? How do we keep our customers?

- **Channels** - How do you communicate with your partners, resources, and customers? How do other companies communicate with them? How are we making this a standard process?

- **Customer Segments** - Is there a specific business line? Is there a specific person in charge of the product or service?

- **Cost Structure** - What are the costs to the model/effort? What activities are most expensive? What resources are most expensive?

- **Revenue Streams** - How do they pay – is there a revenue model? What value are they going to get – what is the ROI?

Business Model Canvas Template

Key Partners	Key Activities	Value Proposition	Customer Relationships	Customer Segments
	Key Resources		Channels	
Cost Structure			Revenue Stream	

When Should I Use It? Why Would I Use It?

When you have an idea or have some form of discovery to pursue. This technique will provide the direction and justification for your effort. It does not mean you need to document everything. It depends on how big your effort is. The goal is to provide clarity and direction. Use the pieces that make sense for what you are trying to accomplish. If you find this model too robust, the next model is half the size and effort of the Business Model Canvas.

Preparation

- Whiteboard or flip chart
- Markers or pens
- Sticky notes
- Spreadsheet on your computer
- Identify your time box for the effort. Shorter for less discovery and longer for more.
- Based on the topics at hand, identify the appropriate individuals to collaborate with on this technique.

How to Get It Done

Always start with the value proposition. It is key to filling out the other sections. Sometimes you will need to reach out to other stakeholders to help with the required information.

Here is an example from the Politics of Pizza House. We have always thought a Pizza Parlor with a political twist would be a hoot. Of course, now someone will carry through with the idea, and we wish them much success!

Key Partners	Key Activities	Value Proposition	Customer Relationships	Customer Segments
- Credit Card Companies - Bank - Landlord - Investors - Suppliers - 3rd Party App Vendor - Phone Providers - FCC	- Sell pizza - Make pizza - Deliver pizza **Key Resources** - Politics of Pizza House team - 3rd Party App Vendor	- Develop a mobile app to sell pizza. - Our target market is mobile app friendly. - They want to order pizza quickly without having to navigate clunky websites, save their preferences and reorder frequently.	- Providing new tech will keep our market and product strong. **Channels** - Daily standups with team - IM with Vendors - No email - Phone as needed	- Politics of Pizza House brand - Gen Y and Gen Z focus - Unusual pizza flavor combos - University students - 5 mile radius from University - Pop up coupon driven buyers
Cost Structure TBD pending research.			**Revenue Stream** TBD pending research.	

Our Challenge to You # 5 – Canvas a Project

Give it a try. Pick a project that you are currently working on or worked on in the past. Try to complete the Business Canvas Model for the project you selected. It doesn't need to be a business-related project. It can be a personal project like painting a room. See how it works for you. Show it to your team, get feedback. Lots of organizations develop their unique customized version (just like you will with Fail Fast and Fail Safe).

Models for Discovery Evaluation – Looking for Personal Improvements

Our favorite for documenting personal improvements is the Agile Vision Board.

Agile Vision Board

The Agile Vision Board (attributed to Roman Pichler) is similar to the Business Model Canvas but is a more straightforward, more focused technique to provide just enough information for smaller efforts. We often get the question "This is a personal thing so why do I have to document it?" Well, how do you hold yourself accountable? The reality is, more often than not, these ideas stay in our heads. They are dreams and not reality. So, if you want it to work, write it down! Dreams are in your head, and goals are on paper. Make yourself accountable.

The template will provide insights you will capture about your effort including:

- **Vision** - A statement or slogan that captures what you want to do.
- **Target Group** - The people that benefit from the effort.
- **Needs** - The main problem that your effort should solve or the benefit it provides.
- **Product** - Three to five features that will stand out or are essential to achieve.
- **Business Goals** - What are your goals? They should be prioritized and quantified.

Agile Vision Board Template

Vision			
Target Group	Needs	Product	Business Goals

When Should I Use It? Why Would I Use It?

Whenever you are stuck and not sure of your career path, your growth opportunities, or what might be getting in your way of achieving success. As mentioned in the Business Model Canvas, this is an excellent alternative if you want a simpler approach.

Preparation

- Whiteboard or flip chart
- Markers or pens
- Sticky notes
- Spreadsheet on your computer
- Identify your time box for the effort. Shorter for less discovery and longer for more.
- Based on the topics at hand, identify the appropriate individuals to collaborate with on this technique.

How to Get it Done

Always start with the vision. Know why you want it, why you are doing it. We usually do the goals next and work our way backward. Fill the columns out in the order that makes the best sense and is comfortable for you.

Vision – Improve my speaking skills. I want to be the "Badass BA!".			
Target Group	**Needs**	**Product**	**Business Goals**
- Coworkers/ teammates on projects - End users who I teach how to use the system - Management that I often present to - Conference attendees that I present to	- More dynamic speaking skills will assist in my career growth - It will provide options for future roles as my speaking skill competencies improve - Provides visibility - More workshop facilitation options	- How to use my body language better - How to modulate my voice when speaking better - How to be more empathetic in my approach - How to feel more comfortable with really large groups	- Get accepted to the big conference this year - Get more involved with local professional speaking organization - Enter the local speaking organization competition

Our Challenge to You #6 – Make a Vision Board

Go write out your own vision board. Put your career objective in the vision box and then complete the model. Make yourself accountable and see results!

3. Engagement Decision

Engagement Decision is where the individual or organization decides whether or not to explore a discovery. Are you going to ignore or be curious about the discovery? If we choose to explore the discovery further, we move into the experiment phase. If we decide to ignore the discovery, we exit the model or wait for the next discovery.

Let's take an example and explore it further. You are walking down the sidewalk and observe the sidewalk has been torn up for construction. A discovery has occurred. The discovery is that the sidewalk is not available for use. Engagement Decision is about making a choice. You can walk on the other side of the street, investigate the reason for the sidewalk being torn up, or both. You might turn around and go back the way you came. Walking away or walking on the other side of the street are tactics to avoid exploring the discovery further. Asking yourself the question "why?" and deciding to find out more about the discovery is a decision to engage.

In another example, you are the product owner for a product. A problem has been discovered with the product when a customer makes a complaint that specific functionality isn't working as expected. As the Product Owner, you evaluate the complaint and perform some investigation. You then decide whether or not you want to engage you and your team to fix the problem with the functionality. This decision is the Engagement Decision.

We exit the Initiation phase when a decision is made to engage the discovery further. We will perform a more detailed analysis on the discovery in the Experiment phase of the Fail Fast model.

The primary outcomes of the Engagement Decision phase are:

- **Agreed upon goal or objective.** The overall goal or objective is established and agreed upon. Typically, this agreement is made with the entire team and the organization's leadership.

- **Agreed upon decision to move forward.** The team and the organization's leadership agree to move forward with the experiments.

Models for Engagement Decision – Corporate Environments

The Force Field Analysis model is one of many useful techniques to use in business situations to facilitate decision making. The model guides you in measuring the positive and negative the factors for a decision. Kurt Lewin created force Field Analysis in the 1940s. Lewin initially used it in his work as a social psychologist. Today, however, it is also used in business, for making go/no-go decisions.

Force Field Analysis

It's typically used to analyze the pressures for and against a decision and has some resemblance to a pros and cons lists. It's a more graphical way to display a pros and cons list but it goes a step further, and it shows us the forces for change and the forces that resist change.

When Should I Use It? Why Would I Use It?

Use the diagram to make decisions. You can also use it to identify resistance or negative points about a decision. By identifying those topics that are against the decision, you take each negative and work to make that negative less significant or remove it entirely from the list.

Preparation

- Whiteboard or flip chart
- Markers or pens
- Sticky notes
- Spreadsheet on your computer
- Identify your time box for the effort. Shorter for less discovery and longer for more.
- Based on the topics at hand, identify the appropriate individuals to collaborate with on this technique.

Force Field Template

Proponents → Decision ← Resistance

Decision

Should we stop paying for cable TV?

Proponents
- Reduces monthly expenses
- We don't have to buy packages deals with channels we don't want.
- We can get rid of the DVR
- All the TV programs we watch are available on Apple TV with a cable subscription

Resistance
- Seems more costly paying for each channel instead of bundling
- We have limited bandwidth on our data plan – it could be slow
- Local TV channels aren't available on Apple TV
- We're not sure it will work on all TVs we have in the house

How to Get it Done

Let's create a table or use the diagram above. The first column will be called "Proponents" and will hold all the positive reasons for saying yes to the decision. Also known as the forces for change. In the middle column write out the decision the group needs to make. This is often called the status quo line. The next column will be called "Resistance" and will hold all the negative items or resistance against the decision. Also known as the forces resistant to change.

We typically start with the proponents for change then move to the resistance. It is harder sometimes to think about why the change needs to happen and easier to be negative about it. Spend more time on the positive. Brainstorm and elicit from the decision makers both positive and negatives about the decision. Validate the points made on both sides to ensure they are not being overstated. Use this to weigh all the factors for making a decision.

Is there a clear winner? Often, you can see that some of the items are "pushing" the other side back. This is the "force field" moving one side or the other. In some cases, there is not a clear winner, so you need to add criteria. You can score each positive or negative force with a score of 1-5. One means it has little significance in the grand scheme of things and five means it holds great significance. 2, 3 and 4 are in the middle. Total the scores on each side. Which side is higher? The side with the highest score points to the decision to make.

Our Challenge to You #7 – Force Field Analysis

Try out force field analysis on a decision you need to make right now. Did it validate or surprise you in any way?

Models for Engagement Decision – Looking for Personal Improvements

Modified Pro and Cons List – Plus/Minus/Impact

The most common way of making a quick decision is the pros and cons list. However, we find that it is not always the best way for making a decision. The reason is that the variables are not 1:1 and they become harder to manage when you weigh positive against negative. The answer to this conundrum is easy. Create a plus/minus list or pros/cons list. Then add a 3rd column called "impact.'

When Should I Use It? Why Would I Use It?

Use this model whenever you need to decide about a personal effort. This model can also be used in business decision making.

Preparation

- Whiteboard or flip chart
- Markers or pens
- Sticky notes
- Spreadsheet on your computers
- Identify your time box for the effort. Shorter for less discovery and longer for more.
- Based on the topics at hand, identify the appropriate individuals to collaborate with on this technique.

How to Get it Done

1. Make a list of the plusses (pros) for why you should do it.

2. Make a list of the minuses (cons) for why you should not do it.

3. Document the impact of the pro and con.

4. Weight each plus with a +1 or a +2 or no score at all (which means neutral). Weight each minus with a -1 or a -2 or no score at all (which means neutral). The plusses and minuses are a gut response to how you feel about the statement. It creates a weighted ranking. You can add more scores if the 1, 2, or neutral scores are not giving you enough variety. Go ahead and add a 3 or 4 into the mix if needed.

5. Total up the scores. If the plusses are more significant, then you should move forward. If the minuses are more significant, then the model indicates the risk is too high or the time is not right.

Modified Pro and Cons List - Example

Question: Should I change jobs to the ABC company?

Plus +	Minus -	Impact
I have industry knowledge. +1	I don't know their processes and systems.

0 - Neutral | If they need subject matter expertise, it will take a while to get up to speed. I have very little extra time right now. Will my performance suffer? -1 |
| The pay is significantly more. +2 | I have an excellent vacation package where I am at, and it won't be as good at the new place. -1 | I don't have extra time because of personal family commitments. I have two children and elderly parents to take care so I may need the extra vacation time. -2 |
| My friend James works there, and it would be great to hang out more. +1 | My boss likes the work I do so I get more flexibility to do what I need to do. -2 | I gain a friend and lose someone who gives me what I need. -1 |

Given the nature of what is happening at home, the money is probably not enough to move right now.

Our Challenge to You #8– Modified Pros and Cons List

Try out the pro and con list. Was it a pro or a con decision?

Initiation completed. Let's see what happens in the experiment phase!

Chapter 6
Fail Fast Model
Part 2 - Experiment

"Success always occurs in private, and failure in full view."

- Proverb

The Experimentation Phase uses David Kolb's ELM (Experiential Learning Model) as its core and builds upon that learning model, to create a model to learn from failure. The model provides the framework on how to fail fast.

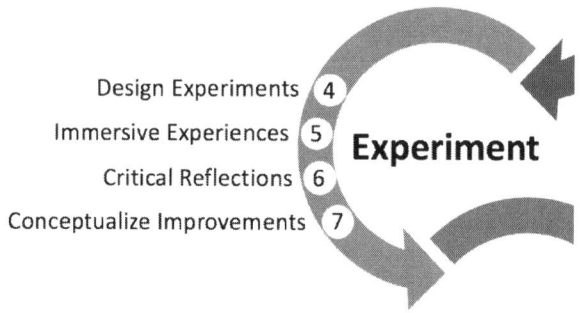

4. Design Experiments

In this part of the Experimentation Phase, we are designing and creating the experiments we will execute in Immersive Experiences. Design Experiments is where we formulate or ideate activities so that we can better interact with the discovery. Design Experiments are also tasks to be considered to troubleshoot or resolve the discovery. Here are some typical tasks that can be performed in Design Experiments:

- Define tasks to learn more about the discovery
- Determine steps to troubleshoot a discovery
- Perform research to understand more details of the discovery
- Reading product manuals
- Meeting with a customer or team members to understand their points of view on the discovery
- Determine the context or how the discovery fits into the global view of the organization or product
- Define the boundaries or scope of the discovery
- Benchmarking the discovery against similar discoveries
- Forming a hypothesis or theory
- Recalling previous experiences with similar discoveries
- Planning the execution of experiments or interactions with the discovery
- Deterring how to test the discovery for further understanding

When designing experiments, ask yourself the question, "What if I…?" Develop what-if scenarios around the discovery. What if we tried to do this with the discovery? What you happen if this switch was flipped? Would adjusting the screen brightness make it easier to read the screen? What if we unplug it while it's running? Would this change fix the problem?

Generating a significant number of experiments is a good thing because it allows you to get creative and to challenge the discovery from many different angles. With lots of experiments comes the realization that not all those experiments can be performed. We need to pick the experiments that provide us with the most insight into the discovery. Once all the experiments are created, the prioritization of these experiments will need to be performed. Prioritization is useful even in cases where there are only a few experiments. A simple comparison of the benefits of each experiment will help put the experiments into a prioritized order.

Typically, people rank priority with high, medium and low. This tends to be ineffective because you get the response that "Everything is a high priority!" No. No, it is not. That kind of thinking tends to ruin progress and create levels of failure that are fast but not safe.

Here is a very simple technique based on a visceral (gut check) response you will have when comparing two items.

Prioritization

I have seen lots of names for this technique including 20/20, Bubble and forced ranking. It is something that we have used long before it had a name.

When Should I Use It? Why Would I Use It?

Use this technique whenever you need to weight out the priorities of multiple items, concerns, requirements or decisions.

Preparation

- Whiteboard of flip chart
- Markers or pens
- Sticky notes
- Spreadsheet on your computer
- Identify your time box for the effort. Shorter for less discovery and longer for more discovery.
- Identify key resources that can corroborate your findings from this technique.

How to Get It Done

Step 1 – Create a list of all the items that you need to prioritize. Write them down on 1 3x5 notecard, sticky note or line.

Step 2 – Mix all of the items up, so you do not have a specific order to them.

Step 3 – Take the first item (Experiment A) and place/write it on the board.

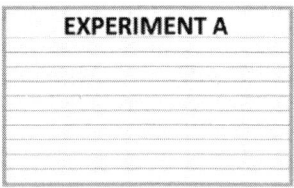

Step 4 – Now take another item (Experiment B) and place it next to Experiment A.

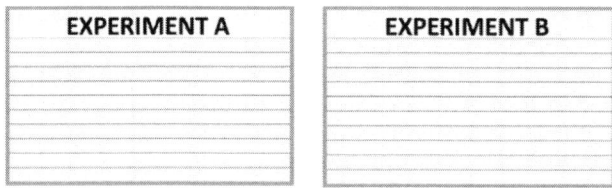

Which is more important? Don't overthink it – go with your gut. If B is more important than A put B on top of A. If B is less important than A put it underneath A.

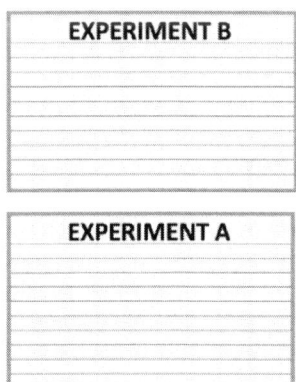

Step 5 – Now compare a new item (Experiment C) to the last card you placed. In this case that would be Experiment B.

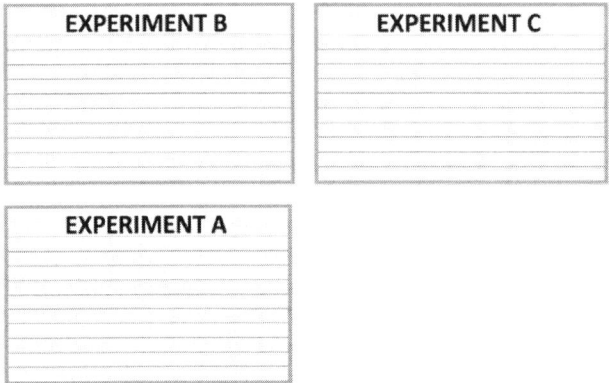

Is Experiment C more or less important than Experiment B? If less, compare it to A.

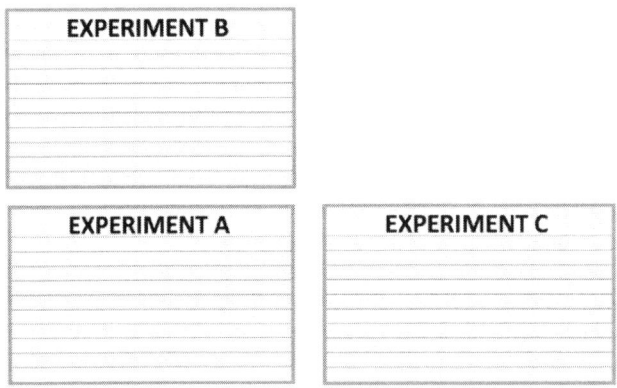

If it is less than Experiment A, then place it underneath.

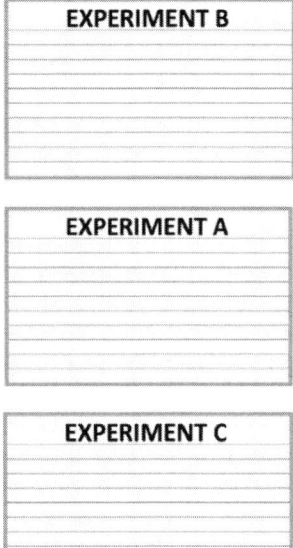

Step 6 – Let's say that you cannot decide if Experiment C is more or less important than Experiment A. Use the following criteria to help determine which is more important and break the tie:

- Business Value
- Compliance
- Technical Complexity
- Reputation
- Dependencies
- Impacts
- Usability
- Cost
- What other factors might break the tie?

If Experiment B has the highest business value, it is likely to be at the top of your list. However, it also depends on what you value. Maybe reputation is more important. Compliance usually wins out over everything.

Our Challenge to You #9 – Prioritize

Use the prioritization method above to prioritize the criteria that you can use to break ties. This will help you be ready for a tie. Go ahead – which is more important? Technical complexity or dependencies? Business value or cost? Knowing the order of these will make it clear when you have ties.

Next, use the prioritization method for fun with your team to get used to the technique. Have everyone write down one item that they would like to have with them while they are stranded on a deserted island with the following exceptions: nothing that requires power, nothing that allows them to escape the island. Anything else goes; alive or not, real or fictional. Have some fun figuring out what is more important. A net for fishing or Bear Grylls wilderness expert?! An unlimited supply of diet coke or wine? A spouse or (insert favorite movie star here). The results should be quite amusing (see the chapter on Fun in Failure and why this will be a good thing)!

A significant part of Design Experiments is understanding the desired or expected outcome. What will determine if we are successful? What are the measurements we will use to ensure we are going to be successful? Defining a successful outcome ensures you know when success has been achieved. If you have no baseline or definition of success, it is hard to tell if you have been successful at experimenting with the discovery.

Assumptions will always come back to haunt you. Take great care to ensure assumptions are validated for all experiments. Additionally, you will need to validate the assumptions you have made about the discovery during the initial phase of Discovery Evaluation. Assumptions can lead you to create experiments to validate those assumptions.

Keep in mind that we are not executing the experiments. That will happen in Immersive Experiences.

The primary outcomes of the Design Experiments phase are:

- **Define what experiments will be executed.** What experiments are we going to accomplish? What is the objective of the experiment? What is it that we hope to learn from experimenting? Who will be conducting the experiments?

- **Prioritize the order in which experiments are to be executed.** What experiments are the most important? What criteria was used to determine which experiments are a higher priority?

- **Define the criteria for a successful outcome.** What is the expected outcome of the experiment? What criteria determines if the experiment is successful?

- **Build play time into your experiments.** Planning experiments is essential but adding some play time or free-for-all experiments can be helpful. Put aside some time to play with the discovery.

Depending on your environment or organization, there might be more things to consider so run through your list of who, what, where, when, why, how and how many.

Think about the "Quantity versus Quality" section you read previously. You want to create a large number of experiments to give you the most interaction with the discovery.

The primary objective of Design Experiments is to plan how you will interact with the discovery. Planning is an essential part of learning and failing fast. Seems counterintuitive to prepare for speed because planning can take a few hours. Having a plan of action will enable you or your organization to move more quickly through the model. Being thoughtful and planning prevents time and resources being wasted just trying something without thinking. It's like planning a vacation. You could get into the car and start driving north. You'll have no idea if your path will take you to something vacation worthy. A little planning around where good vacation spots are located, reservations, and a route calculated on your GPS make the experience far more productive and enjoyable.

Don't create overly complicated plans that are too detailed. Keep the plan high enough that it will guide you down the path of experimentation, but not so detailed that every step by step instruction is planned. Overly detailed plans are not manageable in most cases and hard to follow. Keep it simple and lay out the path explicitly. Don't forget about play time. Interacting with the discovery in a free-form manner will allow you to learn more about the discovery. Set aside some time and a place where you can play with the discovery without executing specific experiments.

We all feel like we need to act immediately, so you look like you are doing something. Thoughtful planning makes you appear like you are doing nothing, but a good plan lays out the easy path to follow. Plans don't always work out, but not having a plan is a guaranty that the next steps in the Fail Fast model will be more difficult, frustrating, and time-consuming than needed.

5. Immersive Experiences

In this part of the Experimentation Phase, we will conduct or execute the planned experiments. The purpose of Immersive Experiences is to experience and interact with the discovery by experimenting with it. By immersing ourselves and utilizing hands-on experiments, we will learn more about the discovery faster.

Interacting directly with the discovery will produce the best results. Feel free to play and interact as much as possible to learn more deeply about the experiment. Remember to permit yourself to fail. We learn best by failure in a safe environment.

In some cases, we are unable to interact with the discovery directly. Using a model, test environment, or prototype is an acceptable way to interact with the discovery. It is best to ensure that when using these techniques, you validate that the representation of the discovery as a model, test environment, or prototype closely matches the real-life discovery as closely as possible. Adam Steltzner understood this concept. How do you test the landing of the Curiosity on Mars? Using computer-generated models. If you are unfamiliar with Adam's story, check him out online. It is a great story where experimentation played a key role in the outcome.

Look back at the Quantity versus Quality section of this book and try to gain experience by interacting with the most amount of quantity avoiding the need to be perfect when interacting with the experiment. Don't try to avoid mistakes but instead try to make them happen. Use your play time with the discovery to test out boundaries and to help develop new theories or experiments.

Frame the Immersive Experience as something positive and fun. Being positive helps you see things more clearly. A negative attitude will create negative results. Approach the failure as a way of learning – not as something wrong or evil.

Think of yourself as a beginner. Don't let yourself or others frame the failure as negative. Failing isn't about intelligence. It's about learning. The reason play time works when interacting with a discovery is that you see yourself as a beginner and you understand that failure will occur. This failure will lead you to learn more about the discovery

Record the results of the experiments for later analysis in Critical Reflection. Keep a log on your experiences with the experiments.

Were the experiments successful? Were your success criteria met? Did you encounter any problems or issues as you executed the experiments? Did you think of something new you would like to try as an experiment in the future? Capturing this type of information will allow you to reflect more fully in the next step of Critical Reflection.

The primary outcomes of the Immersive Experiences of the model are:

- **Conduct or execute the planned experiments.** Get hands-on experience with the product to more fully understand it.

- **Record the results of the experiments.** What happened during the experiment? What went well with the experiment that made it productive? What didn't go well for the experiment? What was confusing about the experiment that didn't make sense?

- **Compare the experiment results against success criteria.** Did the experiment results meet the agreed upon success criteria? Did the experiment fail to meet the success criteria?

Depending on how robust your experiments are, you may want to consider techniques like check sheets, run charts, quality control charts, and Pareto charts.

The Check Sheet

A check sheet is a simple tool for tracking how many times something has occurred or has been used.

When Should I Use It? Why Would I Use It?

Use a check sheet whenever you need to see patterns of use while you are experimenting. This technique can be used in the Critical Reflection section as well.

Preparation

- Whiteboard of flip chart
- Markers or pens
- Sticky notes
- Spreadsheet on your computer
- Identify your time box for the effort. Shorter for less discovery and longer for more discovery.
- Identify key resources that can corroborate your findings from this technique.

What it Looks Like

Error	7-1-16	7-2-16	7-3-16	7-4-16	TOTALS
Claim #	1, 1, 1, 1	1	1, 1, 1,	1, 1, 1, 1, 1, 1, 1, 1	15
ID	1		1		2
First Name	1, 1,	1, 1,	1, 1,	1, 1	8
Last Name		1			1
Address	1,			1, 1, 1	4
Claim Type	1, 1, 1, 1, 1, 1, 1, 1, 1	1, 1, 1, 1, 1, 1, 1, 1, 1	1, 1, 1, 1, 1, 1, 1, 1, 1	1, 1, 1, 1, 1, 1, 1, 1, 1	40
TOTALS	18	14	16	23	68

How to Get It Done

Step 1 – Determine what you are trying to track (represented in the example above by the ERROR column. This can be anything you need to track – as many items you need to track. This could have been how many times did you wake up in the night? Via kids, bathroom break, hunger, neighbors, or the dog.

Step 2 – Determine your timeframe(s) represented by the top row.

Step 3 – Check off how many times it happens.

Whether you are using a check sheet or a quality control sheet (which will track occurrences within an established norm of high/low), this could give you insights into your experiments that might change how you craft them going forward. There are dozens of techniques that can be used to create experiments or track them. Just think about why you would need to do either, both or none.

Our Challenge to You #10 – Track It

Are the experiments that you are planning repetitive by nature? Consider trying out a check sheet. Try new techniques that will enhance your ability to immerse yourself in the experience.

6. Critical Reflection

In this part of the model, we will take our experience and memories of our Immersive experience and reflect on it. The objective is to see patterns we might not have noticed when we were interacting with experiment.

Thoughtfully review what happened during the experiment and the outcome of the experiment to look for patterns. As patterns emerge, those patterns can be used in the Critical Reflections task. These patterns will help us understand the failures when experiments don't work and how we can improve the experiments in the future.

Experiments don't always come to a successful conclusion. When Experiments fail, we need to focus on potential reasons why the experiment failed. Use root cause analysis techniques to dig deeper into the possible reasons why the experiment failed. These reasons will be used later to Conceptualize Improvements.

Use your critical thinking to review the experience, but don't beat yourself up with being overly critical. The same is true with collaborative reflection. Don't allow the group to focus on negative observations that put up roadblocks to learning about the failure.

Groups can get stuck in trying to solve the problem rather than understand it. The purpose of critical reflection is to reflect upon the experience and understand what happened. It's now about solving the problem. That can be hard for some groups to understand because it is our human nature to solve problems. Gaining agreement on the root cause of the failure is an essential step for any collaborative group.

Taking the Criticism

"Any fool can criticize, condemn and complain—and most fools do. But it takes character and self-control to be understanding and forgiving."

- Dale Carnegie

With any failure will come criticism from those around you. It's never comfortable to be criticized. Criticism seems to breed more criticism. Everybody wants to get on that bandwagon and be a critic. It doesn't take much for criticism to start spiraling into a vicious circle.

The only way to avoid negative, or disparaging criticism is to do nothing, but then, of course, you will achieve nothing, and you won't be happy.

"If you never want to be criticized – don't do anything new"

- Jeff Bezos, Amazon.

It is very hard not to get frustrated and defensive when we get criticized. Being defensive can feed criticism to get it spirally out of control. Aggressive disputing criticism will always make that criticism worse. Try to see the criticism as something that is not personal, even when others try to make it personal. Step back and determine if your emotional response is too strong. You may need to step away entirely to give yourself some breathing room to think it over more clearly and understand the criticism.

Here are the steps to take when taking criticism:

1. Don't go with your first reaction. In a split second your emotions will take over and prevent you from hearing the criticism. Check for any signs of anger and frustration. It's not time for your emotions to process the situation, it's time for your brain to handle the situation. This is very difficult. Our facial expressions and emotions respond much faster than logical thought. Although it is ideal to be expressionless when getting criticism, it's a difficult thing to accomplish. It's an even more significant challenge to accept criticism from individuals you don't respect. Hold your cool. Return your face to expressionless.

2. Criticism can be beneficial once you logically think through that criticism. Not all feedback is valuable. Listen and try to understand the criticism without getting defensive. Even flawed perceptions and feedback can be made useful if you logically review them.

3. Listen carefully and try to understand. See the emotional state of the person providing the feedback and be careful not to allow your emotions to escalate. Allow them to share their thoughts

entirely without interruption. When they are done, repeat back to them what you heard. Don't assess their statements at this point. Recognize the person may have difficulty giving feedback and isn't communicating it well.

4. This is the hard part. Say thank you for the feedback.

5. Now ask questions to deconstruct the feedback. Get specific examples. If no examples are forthcoming, let them know it's hard to take action on something that isn't understood. "Please understand it's difficult to understand your feedback when you can't elaborate specific examples. I'm trying to take action based on what you said, but it is difficult without specific examples."

6. If the feedback is correct, then acknowledge that the feedback is accurate. "You are correct. I excluded him from those meetings. In a private meeting, I apologized and brought him up to speed." The person giving feedback may not be aware of what steps you took behind the scenes.

7. Ask for concrete solutions. "How would you have handled this situation? When you handled situations like this what made it successful?"

8. On more substantial issues that can't be resolved with the meeting time, propose a follow-up time. If you need time to think about the feedback, ask for it. Think about the feedback logically.

Once the criticism is more fully understood, choose to take or leave it. If the criticism is valid, then accept and learn from it. If the criticism is junk, say thank you and move on. No further discussions are required.

Handling criticism and taking feedback on your performance and your experiments is an essential part of failing safely. You will get lots of criticism and feedback on the outcomes of your experiments in the Immersive Experiences, Critical Reflections and Conceptualize Improvements phases of the model.

Critical Reflection as a Team

Before a collaborative approach to Critical Reflection can begin, ground rules need to be set for how the group will behave and manage itself. Level set the group expectations and specifically call out how the group will handle adversarial or overly critical scenarios. Another essential factor to consider is how to give everyone in the group equal time in expressing their reflections and observations. Equally important to making a collaborative approach to Critical Reflection work is not allowing one or a few members of the group dominate the entire group. Create rules that will allow everyone to be heard and acknowledged positively.

Observations and reflections are the pieces of the puzzle that will fit together to understand why the experiment failed. The broader the perspective, the more likely it is you will see the larger picture. Seeing the outcome of the experiment from many different angles and perspectives allows patterns that emerge to help form the root cause of the experiment failure.

In some cases, the root cause may not be able to be determined. In these cases, you will need to modify or break the experiments into smaller parts to locate the root cause of the experiment failure. If you don't understand why the failure occurred and the root cause of the failure, it's difficult to see how to learn from that failure. Check to see if your experiment design wasn't thought through enough or designed well enough to fail in such a manner that you could learn from that failure. The experiment might be too complicated and not yield the details needed to get to the root cause. You might find yourself being not able to understand the failure occurred. Don't worry that is quite common. The recovery to not understanding the failure is to rethink the experiment to produce more detailed and meaningful results.

Don't lose track of those lessons learned or root causes. Forgetting about the root cause of failure will cause you to repeat the same failure over and over again.

Taking Criticism as a Team

Most often a team is put together to experiment, and that team will get feedback on the outcomes of the experiments as individuals which they pass on to the team. As a team, you will need to accept that criticism, understand it, and respond to it positively.

Teams need a way of requesting and gathering feedback. Having a central place where feedback is registered is an integral part of ensuring the team can not only get feedback from each other but act on the feedback. Agile retrospectives or post mortems are great ways of getting feedback. Consistently keep asking the team for their feedback on experiments and their ideas for improving them. Keep learning from what went wrong and how we could do it better next time. When the team comes up with an idea for the improvement, be sure to take action whether that is to prioritize it against a list of other ideas previously gathered, or to act upon it immediately. Don't let good ideas sit around and never get acted upon.

Stop the blamestorm. It's easy to blame an individual on the team or outside of the team. Blamestorming happens when it is easier to find someone to blame rather than to find a solution or lessons learned from the failure. Learning together as a team is a much stronger option in responding to failure than trying to figure out the next scapegoat. You will be far more productive learning from the failure than trying to throw someone under a bus. It's human nature to blamestorm and important to recognize when it is happening and confront it. Is blaming that person going to solve our problem quickly? Are we wasting time trying to find a scapegoat rather than focusing on a solution?

Being Pleasantly Skeptical

Every experiment needs a healthy dose of skepticism, but skeptical can be a highly-charged word. Are you skeptical? In business being skeptical is akin to being a nay-sayer, challenging to work with, not being a team player or a troublemaker. We disagree. You need skeptics on your teams. Skeptical does not need to be destructive or harmful. You do not have to be a jerk to be skeptical. Skeptical can mean looking at things with a critical eye respectfully and politely to create thoughtful and meaningful discussions to elicit root causes and improved observations of experiments.

Being skeptical is an essential part of interacting with an experiment and the results. It helps you see if the intended result is meeting expectations and the success criteria. Being skeptical also assists you in making changes to the experiment to get better outcomes.

Baloney Detection Kit

Carl Sagan (November 9, 1934–December 20, 1996) was an American astronomer, cosmologist, and astrophysicist who was known as a master of the balance between skepticism and openness. In Carl's final book, he wrote a chapter titled "The Fine Art of Baloney Detection" where he outlines the "Baloney Detection Kit." When a new idea was introduced, Carl Sagan used the kit to evaluate them. If the new idea survives examination by the tools in the kit, the idea was given a warm reception and sometimes a provisional acceptance of the idea. If the new idea was challenged by the kit and didn't fare well, the idea was refined and modified.

The Baloney Detection Kit is a collection of tools that can be applied when you are experimenting. Although these tools were designed for the broader realm of cosmic exploration, physics, and science, they are equally at home with more earthly experiments.

Experiments are used to prove out a hypothesis. A hypothesis is a potential or unproven idea that could solve a problem or issue. You can have more than one way to solve a problem or issue. Although hypothesis and experiment are terms commonly used in science, they can be used in business. Here's the baloney detection kit:

1. Wherever Possible There Must Be Independent Confirmation of The Facts

We have learned several hard lessons in dealing with Commercial Off-the-Shelf (COTS) or Out-of-the-Box Software (OOBS) vendors. The mistake was accepting the facts at face value and not verifying those facts.

For the COTS project, it meant accepting the vendor's statement that their product would perform the needed functionality without verifying the desired function worked. Building a sandbox to use as a playground to confirm the desired functionality worked in the COTS application would have stopped us attempting to implement that solution. We would have moved on to another solution.

Don't be swayed by the salesman pumping up all the good things about their product when a simple test could show it's a lemon. Look at the experiment results with a skeptical and realistic eye. Dig in a little deeper and verify the results of the experiment. If you get asked to skip the experiment, that should raise a red flag. Without hands-on interacting and performing designed experiments.

Keeping balance here is important. Don't focus entirely on the good or bad. Observe more globally to ensure your critical reflections don't get overly positive or negative.

Involving your team is very helpful to ensure you are getting a broader perspective of the results. Understand their point of views as a way of enhancing your understanding of the experiments and outcomes. Achieving the 360-degree view of the experiments and their results will provide you with significant insight. This insight, in turn, will help you conceptualize improvements to the experiments in the next step and move you faster through the model.

To gain an even broader and independent view of the experiments and their results, review your experiments and results with others outside of your team or organization. You are looking for an unbiased review. Even discussing the experiments with your customers can provide broader insight. Such external reviews can take time to coordinate, but they are well worth the effort. External reviews can be as simple as having a review of the experiment and results or recreating your experiment independently and trying to get the same results you got in your experiments.

If both your experiment and the external experiment get the same results that will help validate your results as being unbiased. If the external experiments get differing results, that's a flag that the experiments are not repeatable, and the results may not be impartial or accurate.

2. Encourage Substantive Debate by Knowledgeable Proponents of All Points of View

A forest is seen in many ways. From above you see the canopy of green leaves. From the ground, you see the tree trunks and branches. Looking from below ground, you see the roots of the trees. Different perspectives give you a deeper understanding of the overall situation that leads to more efficient and meaningful solutions. Having a support system in place and others that are curious like you can help provide that broader viewpoint.

When working with a client who had three different locations for their labs Paul was asked to create a single process and join all their labs together under one big tent called enterprise product testing laboratories. The labs mostly did not talk with each other because they were in different parts of the world. The project was to upgrade a Laboratory Information Management System (LIMS) in one of the locations because the vendor no longer supported the version the lab was working on and had a lack of connection to all their new lab equipment. The lab manager assured me that talking with the other labs would be a waste of time. My team started looking more globally inside and outside of the organization. My team spotted an article on the web about one of the organizations other labs upgrading to a new system. We reached out to them to get a better understanding of their tool and the effort that would be required.

Surprisingly we discovered they had a tool that would meet all our needs, an enterprise license for the software that would save our project money and desired a better connection to our lab. The overall result was the reduction of the project budget and project duration cut in half. We got lucky in this situation, but that luck was created by looking more globally across the entire organization and even externally to the organization.

Encourage debate with others that are knowledgeable about your product or experiment. Having a debate with individuals that are unfamiliar with the product or experiment can lead to frustration. If the other party isn't fully aware or lacks knowledge, significant time is spent bringing them up to speed on simple things that experts in your area would have known. Experts will be able to provide you with a greater insight into the possible reason why the experiment failed and the interpretation of the experiment results.

3. Arguments from Experts Carry Little Weight

This tool is the most charged of the tools in the kit. It almost seems to invalidate the previous rule. The statement here is that no one is all knowing or god-like in his or her knowledge of a subject, system, or process.

Engaging experts does have great value, but experts can be biased. Experts can make assumptions based on their experience. You will need to handle these expert assumptions with great care. As with all assumptions, verification is required of those assumptions. Trust but verify. The world doesn't sit still. Experience is the most valuable when the expert is adaptable to an ever-changing landscape of knowledge.

Challenging an expert on their knowledge in a meaningful and positive way is vital to ensure that the expert is seeing the entire global picture as it exists currently not how it existed in the past. Challenging doesn't have to be a negative experience. You are asking the expert to help you understand their thought process. A bad situation can happen when the all-knowing expert doesn't like to be challenged. These experts begin to "spin" the facts to avoid looking like they made a mistake and are intense in their desire to maintain the status of all-knowing. They do not want to lose face.

Be aware of when an expert if being genuine in helping you to understand their viewpoint. They should be asking questions of you as you ask questions of them. It should feel more like a conversation. You should feel you can trust their judgment. In your discussions, there is a give and take. Experts that don't seek to understand your position but to enforce their own instead are of little value. Don't be bullied to accept their viewpoint until you can fully understand it and verify it.

Seek to understand first then challenge appropriately. Get a second opinion on the viewpoint where you can and verify the expert's views with other experts or your own experiences.

4. Spin More Than One Hypothesis

When we have a problem or an experiment, we tend to focus on one solution, cause or potential outcome. Many times, we don't create all the possible ways into which an issue could be addressed and weight them against each other. By focusing on the first solution, you are ignoring the potential that other better solutions are available and will work. In your experiments, try to figure out alternative experiments that would produce the same result or resolve the issue.

Think about when you are buying a house. You don't typically buy the first house you see when searching for a new home. You see lots of homes that are available and weigh those choices against each other.

In business, we feel we must choose only one path and not look at other potential opportunities. This closes us off to find the best outcome for the experiment or issue. Weighing solutions against each other have an unexpected benefit of combining bits and pieces of different solutions to form a new solution. This cross-pollination of ideas creates interesting new solutions. When looking at a list of solutions, see if taking parts of one solution and combining it to make a better solution. We sometimes confine ourselves to a single application or product when combining multiple applications, processes, and product might create a better solution. Be creative in creating solutions and try to avoid limiting your thinking.

Back to the house buying scenario. As you walk through each of the potential choices, you discover there are things you like or dislike about each possible house. These likes and dislikes help you form a better picture of what will suit your needs in a new home. In your head – or better yet write them down – you keep track of these likes and dislikes. Viewing these potential candidates shapes your view of a successful outcome. You select a house that is best going to meet your like and dislike criteria. Even if the house purchased is missing something you liked in another house, you will keep that criterion in your head for later remodeling and upgrades.

If there's something to be explained, think of all the different explanations. Think of tests by which you might systematically disprove each of the alternatives. What survives has a much better chance of being a practical solution than if you had only run with the first idea.

5. Try Not to Get Attached to a Hypothesis Just Because It is Yours

A better way to state this would don't get hooked on a solution you came up with and ignore other potential solutions. Be open to other possible business solutions and be open to modifications or changes to the business solution you have put forward. Building the best business solution needs flexibility and collaboration.

Ask yourself why you like the idea. Compare it fairly with the alternatives. See if you can find reasons for rejecting it or modify the solution to strengthen it.

6. Quantify

If whatever it is you are explaining has some measure, some numerical quantity attached to it, you will be much better off measuring each potential solution to discriminate among competing solutions. Quantification solutions are important. Ensure analysis can quantify the solution. Figure out the measurements that are the most meaningful and apply them. Every solution should have an objective measurement that can be used to determine its value when compared to other solutions.

Anything that is vague and open to many different explanations will result in solutions being confusing, frustrating, and easily misinterpreted. Always be aware of the assumptions you have made that are part of the solution. Verify your assumptions as quickly as possible. A lousy assumption can lead you to a solution that isn't valid or appropriate for your needs.

7. Every Chain Link Must Work

If there's a chain of arguments or decisions made for every solution, then every link in the chain must work (including the premise) — not just most of them. Solutions are combinations of many things like interfaces (systems, hardware, software, people), processes, data, interactions, and business rules that are all chained together. Verification of the links in a chain is called traceability.

For example, the premise is that the electrical outlet needs to be replaced to solve the problem of the floor lamp not working. The chain of arguments is that the electrical socket doesn't have any power by measuring it with an electrical meter. When the lamp is plugged into a different outlet, it works just fine. Replacing the socket is the opinion and suggested solution of your brother-in-law the electrician. He has seen this type of thing before. Further investigation revealed the fuse was not blown.

Sounds like a good chain? What about the switch on the wall that controls the socket? Was that tested, or did we assume it was working? Did we make sure the meter measuring electricity in the socket was working? Verify everything in the chain of thinking to ensure you didn't miss anything, and it makes sense. Verify that all potential solutions are included in the chain.

The result of our example was that the wall switch that controls the outlet was not working.

8. Occam's Razor

Occam's Razor is the problem-solving principle that the simplest solution tends to be the right solution. When faced with two hypotheses or solutions that solve a problem equally well, it is wiser to choose the simpler or less complicated solution. Simple solutions are more straightforward to implement and easier to understand. The most significant challenge is taking complex business problems and solving them with simple business solutions. Strive to eliminate the complexity of your solutions to keep them simple. By creating solutions that can quickly be implemented, we are then able to move on to other solutions needed for the product. Want to choose the simpler solution? Choose the one with the fewest assumptions.

9. Always Ask Whether the Hypothesis Can Be, At Least In Principle, Falsified.

This tool is the second most charged tool in the kit because of the word falsified which has a substantial adverse reaction. Undergo a careful and thoughtful review of the outcomes to ensure that positive or negative results are verified. Look for bias in your experiment results. Verify to ensure the data used to support the experiment results is valid and relevant.

The best way to do this is to open up your experiments and their outcomes to critical feedback from others. This opens up collaboration with others who are invested in the solution. Open experiments or prototypes to others so they will be able to interact with them to see if they get the same results.

The primary outcomes of the Critical Reflections phase of the model are:

- **Pattern recognition.** Did any patterns emerge when reviewing the experiment results? Looking at the results of all the experiments, did any global patterns emerge? Is there a potential root cause that is affecting all experiments?

- **Potential failure causes.** What caused the experiment to fail? Could a different experiment yield better results? Where could experiments be changed to make them more effective?

- **Acquire feedback.** Was feedback on the experiment or experiment results obtained? Was the feedback provided from an unbiased source? Was the feedback offered by someone who is knowledgeable about the experiment? Did you review the experiment results with external, unbiased, and knowledgeable resources?

- **Baloney Detection Kit.** Did you apply the kit to the experiment and its results? Did you apply the kit to feedback on the experiment? Was the data used to support the experiment results verified?

- **Result verification.** Could other replicate your experiment and get the same results? Was the data used to support the experiment results verified as accurate?

7. Conceptualize Improvements

Here we are thinking about potential new ways in which to avoid the root cause of failure. We are not designing the experiment. We are determining what the experiments will be designed to achieve. We are taking the critical reflections from the previous step and turning them into potential solutions.

It is possible to arrive at the Conceptualize Improvements part of the model without having to make any improvements. The experiments we previously conducted verified the problem didn't exist. Additionally, a solution that was designed and implemented that solved the problem. If no further solutions are needed, then no additional work is necessary, and we would move directly into Viability Analysis.

If our previously conducted experiments indicate a solution is still needed, we will find root-causes and determine potential solutions. The goal is to take our critical reflections and turn them into ideas on how that failure could be avoided in the future or how a solution could be applied to eliminate the failure. If the failure was due to a specific metal not being able to support the weight of the product under certain circumstances, we would then think of other potential materials that could be used. We are creating a hypothesis. In Design Experiments, we could build the experiments or tests that we would use to prove out our hypothesis.

We have discovered a potential flaw in the design of our product because the metal used might not support the full weight of the product. We designed experiments to test this discovery out in Design Experiments. We executed those experiments in Immersive Experiences. We critically reflected on the results of the experiments in Critical Reflection.

Now when we are in the Conceptualize Improvements part of the model, we are building a list of experiments we believe will solve the problem of the metal not supporting the weight of the product. At this point, we are looking to see if other metals will work better. We review the design to see if the thickness of the metal will correct the problem. We think up many other potential ways in which the discovery can be resolved.

Now we have a list of potential options for experiments. We now need to evaluate the possible experiments against each other for prioritization.

Wait – didn't we prioritize in Design Experiments already? Why are we doing it again?

In the Design Experiments phase, we prioritize the order in which we want to execute the experiments. In the Conceptualize Improvements phase, we are prioritizing the list of potential options we want to explore during the next cycle of Design Experiments, Immersive Experiences, and Critical Reflections. Multiple potential options can be associated with each experiment. That list of potential options can get overwhelming, and we need to prioritize them to focus on those potential options that will provide us with a higher number of successful experiments.

The primary outcomes of the Conceptualize Improvements phase of the model are:

- **Determine the root cause(s).** If an experiment failed, was a root cause found? Would the root cause of a failed experiment cause other experiments to fail or succeed? If the experiment failed, why did it fail? Can you validate the root cause is correct?

- **Create potential experiments.** Does the potential experiment address a root cause? Does the potential experiment align or support the agreed upon success criteria?

- **Prioritize potential experiments.** Did the potential experiments get compared to each other to obtain an agreed upon prioritized list of experiments? Was it determined which potential experiments will most likely be worked on? Is there agreement on the prioritized list?

- **Validate the success criteria.** Does success criteria need to change because we understand the product and the rationale for the failed experiments more fully? Can we realistically achieve the agreed upon success criteria? If we change the success criteria, does that invalidate potential experiments or cause experiments to be re-prioritized?

Here is where you need to be careful. How many times are you going to do this? Lather. Rinse, Repeat. If you have all the time and money in the world that is one thing but getting trapped in an endless loop of experimentation is another. Prioritize and move on.

Experiments are done. It is time to realize our results.

Chapter 7

Fail Fast Model Realization Phase and Model Walk Through

"Success doesn't teach as many lessons as failure does."

- Jay Samit

The primary objective of the realization phase is to determine if there is a desire to continue designing experiments or if the experiments should stop and the model concludes. If we decide to continue, we will move back to step 3. Design Experiments in the model. If we choose not to continue, we proceed to step 8. Conclusion.

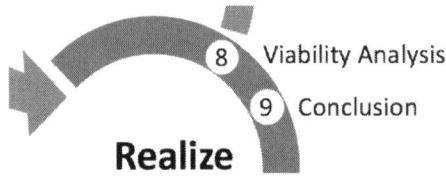

8. Viability Analysis

Viability Analysis is primarily about deciding the products next steps. Do we move the product into engineering? Should it go to production?

To best make those decisions we need to understand the outcomes of all the experiments fully. Were experiments missed? Are the results of an experiment valid? Being skeptical of the experiments and results ensures that the quality of the product or service is not comprised by a perspective that is too focused. Think of it as making sure you are seeing the global picture in which the product or service will exist and verifying the changes you made in experimentation will fit into that global world.

Viability Analysis is the determination of whether or not we have the will or desire to continue experimenting or if we will close and conclude the model.

It is possible that even though we have a prioritized list of potential experiments to design, there is no drive or will to continue to move forward with more experimenting and improvements. We must look at the law of diminishing returns. We will decide at this point if things are good enough. Obtain consent to move into designing more experiments. The individual or organization may need to focus on other priorities. If it is the case that no further experiments need to be developed or conducted, we will move to the conclusion step of the model.

We highly recommend keeping all your experiments and data. They can be used at some future point if it is decided to move forward again with experiments.

If the decision is to move forward, then we will move back to the Design Experiments step of the model.

The primary outcomes of the Viability Analysis phase of the model are:

- **Perform Viability Analysis.** Is there value in continuing experiments? Should we continue to design and execute experiments? Have we met our success criteria fully? Does the individual or organization still see value in continuing the experiments?

- **Prepare for the conclusion.** If we decide not to continue, have we archived our experiments and results so that others can use them in the future? Have we shared our findings and success with others so that they can learn from our experiments?

Consider at this point a decision model to help make a conclusion. Weighted ranking tables, decision trees, decision tables, and root cause analysis techniques could all be useful in making your decision.

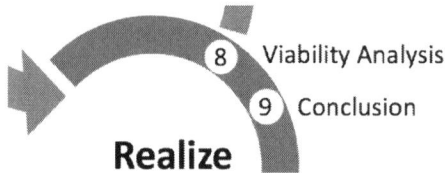

9. Conclusion

Once the decision is made to no longer perform further experiments, either because we were successful in the experiments or there is no longer a desire to conduct new experiments, the team and organization's leadership reflect on the journey in conducting the experiments and collecting results. Looking at the model as a whole, what did we learn? How can we do it faster next time? What should never be done again?

In the conclusion phase, we look back on the model itself and how well it worked or didn't work. Further analysis is performed on the model to see if changes can be made to utilize it more effectively.

We have been through the model, now let's walk through it with an example from start to finish.

Walking Through the Fail Fast Model

"Trying is the first step towards failure."

- Homer Simpson

Model Walk Through – The Toaster Scenario

Let's walk through a scenario where we are making improvements to our toaster product line. We will start with Initiation.

1. New Discovery

Our customers have presented us with a new opportunity to bring a new product to the market. Our current toaster is getting a lot of feedback indicating the design and weight of the toaster is not meeting the expectations of our customers.

2. Discovery Evaluation

We followed up on previously received complaints about our toaster from potential customers. After further conversations with potential customers, we discovered that our customers are looking for toasters that are lighter in weight, smaller and more compact to fit in smaller kitchens, and would toast faster. As a company, we believe that we are a leader in small appliances and would wish to continue that market lead by creating a new innovative toaster product.

The output of this phase of the model is:
- Problem or opportunity statement is created.
- Establish high-level goals and objectives.
- Providing enough information to decide.

3. Engagement Decision

After careful review with our organization's leadership, we decide to move forward on the new innovative toaster project. Our goal is to build a new type of toaster for the home. This new toaster is made of new materials that are lighter and allow us to design a more compact toaster that fits into smaller kitchens. It also uses a new heating mechanism that will toast bread using less energy in less time. Our goal is to make toast fast, energy efficient, and uses a small amount of space on the kitchen countertop.

We set our success criteria concerning weight, color, energy use, toasting speed, and scratch resistance. The organization's leadership agrees and supports the success criteria we established.

Overall Toaster Product Goals & Objectives

1. New material visually matches our other products
    a. The material color matches stainless steel grey #FS016
    b. The material color matches fire engine red #FR117

2. New material is scratch resistant
    a. Won't scratch when rubbed against other stainless-steel surfaces,
    b. Won't scratch with household scrubbing pads

3. New material is light
    a. The overall weight of one cubic foot of the new material is 25 - 50% less in weight than one cubic foot of the old material.
    b. The new material, when formed into an 8-inch width x 8-inch height x 8-inch length box, does not weight more than stainless steel formed into the same size box dimensions using the same construction techniques.

4. For brevity, we will skip over all the of the goals and outcomes outlined in the example above. There are quite a few more that could be derived.

The output of this phase of the model is:
- Agreed upon goal or objective
- Agreed upon decision to move forward

4. Design Experiments

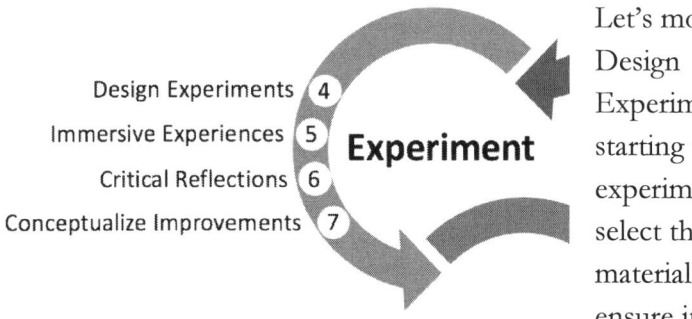

Let's move into Design Experiments by starting to design experiments to select the new material and ensure it will meet our success criteria. Our first set of experiments are focused on making sure the color of the new material is visually compared to the color of the old material. We want the colors to match.

We elaborate on the steps and process needed for the experiment to demonstrate the same process can be used for both the new and old metals. We are using the same process that is currently used in production. We are trying to verify the process currently used will give us a new metal that is painted with the colored heat proof coating will match the current painted metal.

For this example, let's look only at just the first new capability and success criteria.

Our goals and objectives for the experiment are:
1. the material color matches stainless steel grey #FS016
2. the material color matches fire engine red #FR117

To meet our goals and objectives criteria, we created these experiments:

Experiment 1: Material color matches stainless steel grey #FS016

    a. 5 grams of heat proof coating is mixed with 1 gram of #FS016 color
    b. The mixture is applied to new material
    c. The coating is allowed to dry for 24 hours in the drying room
    d. Compare the coated material with our current material and visually compare them to see that both materials are the same in color

Experiment 2: Material color matches fire engine red #FR117,

    a. 5 grams heat proof coating is mixed with 1 gram of #FR117 color
    b. The mixture is applied to new material
    c. The coating is allowed to dry for 24 hours in the drying room
    d. compare the coated material with our current material and visually compare them to see that both materials are the same in color

A plan has been constructed to focus our testing, and now we can move on to the Immersive Experiences tasks.

We prioritized the experiments based on which color is the best-selling. Experiment 2 was ranked the highest priority because fire engine red is our best-selling color. Experiment 1 was ranked low priority.

We defined our success criteria as both colors will match with our other product lines. We will take samples of each color on the existing metal and use that as a baseline to compare the color with the new material. We will visually examine the old metal and the new metals colors to ensure they match.

The primary outcomes of the Design Experiments phase are:

- Define what experiments will be executed.
- Prioritize the order in which experiments are executed.
- Define the criteria for a successful outcome.
- Build play time into your experiments.

5. Immersive Experiences

Experiments are conducted to see if we can meet our success criteria. We performed those experiments we planned. Here are the results:

1. the material color matches stainless steel grey #FS016,
    a. FAILED – color did not match, after application of the coating the material turned green.

2. the material color matches fire engine red #FR117,
    a. SUCCESS – color matched

The primary outcomes of the Immersive Experiences of the model are:

- Conduct or execute the planned experiments.
- Record the results of the experiments.
- Compare the experiment results against success criteria.

6. Critical Reflections

Critical reflection is primarily about determining the root cause of the experiment failure. In this example, the reason could either be the coating itself interacting with the new metal or heatproof coating.

In critical reflection, we use root cause analysis tools like "5 Whys?" or the fishbone diagram to help us understand the root cause of why the experiment failed. Although it's not possible in all cases to know the exact reason for an experiment failing, getting down to the most detailed explanation is essential as this information is used to create new experiments. Getting down to the root cause of the failure will ensure that the next round of experiments will be more successful.

If the experiment was successful, critical reflections are not needed. Conceptualize Improvements can be skipped for the experiment. However, it may be beneficial to understand why one experiment was a success and the other was not. Comparing the experiments might reveal differences and potential future failures. You may have just been lucky with one experiment being a success and the other failing.

After thoughtful analysis, we have determined the most likely cause for the failure was the heatproof coating interacting with the paint color. We compared the experiments and discovered the metal and heat proof coating used in both experiments was the same. We also confirmed the process by which the color was mixed and dried was also the same. Because the process for both experiments was the same, this led us to see the color itself as being the issue. This is an oversimplified example, and there could be many more causes with further analysis.

The primary outcomes of the Critical Reflections phase of the model are:

- **Pattern Recognition** – look for patterns in the experiment results.

- **Potential Failure Causes** – identify any potential causes that caused the failure. Verify the experiments were carried out step by step in the same manner if comparisons are being made.

- **Acquire Feedback** – review your experiments and findings with others to determine if your approach and results are valid.

- **Baloney Detection Kit** – think skeptically about the results of the experiment.

- **Result Verification** – verify your results have valid data points.

7. Conceptualize Improvements

Conceptualization is primarily about the creation of approaches that could solve the problem and correct the cause of the failure.

After getting a good understanding of the cause for the failure, it's time to conceptualize potential methods that would achieve a successful outcome for the experiment. You would not need to conceptualize improvements if the experiment results were successful in meeting your success criteria.

Success criteria can change as you experiment. Loosely defined success criteria can cause a significant amount of churn. Experiments could wind up in an endless loop of trying to meet success criteria if that success criterion was not agreed upon. Carefully manage the change to the success criteria. If the success criteria must change, it is a good practice to ensure everyone concurs with that change before measuring yourself against the new success criteria.

A decision was made to try out three different heatproof coatings that might work better on the new material. This way we can verify that the coating we are currently using is causing the experiment to fail. The conceptualization is that trying different coatings will prevent the color of the new material from not matching our current material and color.

We haven't fixed the root cause yet. We will be going back to design experiments. We will take the information from conceptualization and create experiments to prove out whether or not our ideas for correcting the problem will be successful.

The team will move on to realization once all our experiments are successful in meeting the success criteria. For this example, we will state all experiments are a success, and we met all the success criteria.

The primary outcomes of the Conceptualize Improvements phase of the model are:
- Determine the root cause(s)
- Create potential experiments
- Prioritize potential experiments
- Validate the success criteria

8. Viability Analysis

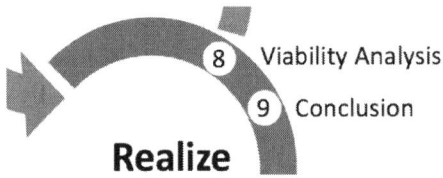

Viability analysis is about reviewing the outcome of all experiments and their relationship to each other. A decision is made whether or not the product design is ready for production design or further development.

After all, the experimentation is completed, and the success criteria are met, the team decides if the new metal should be used in the production of the new toaster.

The primary outcomes of the Viability Analysis phase of the model are:

- Perform Viability Analysis
- Prepare for conclusion

9. Conclusion

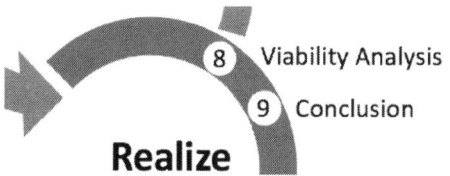

Once the decision is made to no longer perform further experiments either because we were successful in the experiments or there is no longer a desire to conduct further experiments, the team and organization's leadership should reflect on the journey in conducting the experiments and collecting results. Looking at the model as a whole, what did we learn? How can we do it faster next time? What should never be done again?

Model Walk Through – New Reporting Scenario

1. New Discovery

Our Sales team is looking for a new sales report to locate potential upsells for existing customers. Our customers have purchased our base product and now we would like to sell our enhanced services to those customers.

2. Discovery Evaluation

We chatted with the sales and marketing teams to determine the potential increase in sales we could expect for the report. We quantified the business value in potential sales. A high-level goal of increasing enhanced services sales by 17% was established. Since our company has recently launched our enhanced services business, it is a company goal to increase enhanced services sales by 25% this year. The report doesn't fully get us to increasing sales 25%, but it does make a significant contribution. To be effective the new report will require new processes for the sales team to use the report effectively.

Information has been packaged for review by senior leadership to approve the reporting project.

The output of this phase of the model is:
- Problem or opportunity statement is created.
- Establish high-level goals and objectives.
- Providing enough information to decide.

3. Engagement Decision

After careful review with our organization's leadership, we decide to move forward on the report. Our goal is to build the new report and put process in place to help the sales team to utilize the report effectively. We established the success criteria below:

Enhanced Services Sales Goals & Objectives

5. Establish new sales report for enhanced services
6. Combine data from System A and System B to identify potential enhanced services customers
7. Establish new sales team processes to utilize the report
8. For brevity, we will skip over the entire list of goals and objectives.

The output of this phase of the model is:

- Agreed upon goal or objective
- Agreed upon decision to move forward

4. Design Experiments

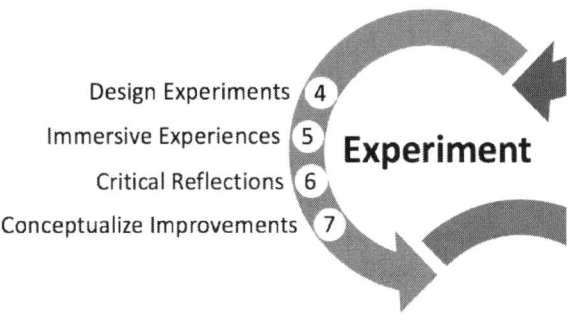

For this example we are going to focus on the data for the report that will need to come from System A and System B. The data from these two systems have not been joined together previously. Let's starting to design experiments to create the new report and ensure it will meet our success criteria. Our first set of experiments will ensure the data that is loading into our reporting system matches the transactional data. We will then need to take that data and decide how to create a formula to combine the data in a meaningful way.

Our goals and objectives for the experiment are:
3. Data from System A matches the data in our reporting system. All data transformations are validated.
4. Data from System B matches the data in our reporting system. We didn't transform that data at all during the transfer process to our reporting system so no further verification is required.
5. Determine the correct formula for customer identification for upsell opportunities.

To meet our goals and objectives criteria, we created these experiments:

Experiment 1: Compare the data from System A to the reporting system. We will be performing an extract, transform and load of data from System A.

2. The extract from System A must match the input to the transformation process.
3. Perform the transform process on the data. Validate the data that is to be loaded into the reporting system has been transformed correctly.
4. Ensure the transformed data matches the data the is loaded into the reporting system.

Experiment 2: Data extracted from System B matches the data in the reporting system.

Experiment 3: The business has devised 3 formulas to identify potential customers for enhanced services. We need to validate which of the formulas is effective and runs the most efficiently (it's not a SQL hog and the query runs fast).

A plan has been constructed to focus our testing, and now we can move on to the Immersive Experiences tasks.

We prioritized the experiments based on which color is the best-selling. Experiment 2 was ranked the highest priority because there is more data in System B then System A. Experiment 1 was ranked next in priority. Experiment 3 was ranked last in priority because we can't calculate the formulas without good data.

The primary outcomes of the Design Experiments phase are:
- Define what experiments will be executed.
- Prioritize the order in which experiments are executed.
- Define the criteria for a successful outcome.
- Build play time into your experiments.

5. Immersive Experiences

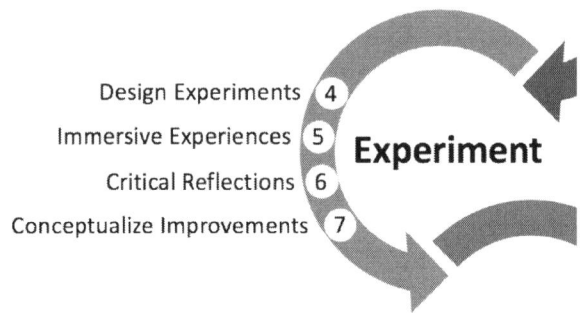

Experiments are conducted to see if we can meet our success criteria.

Here are the results:

- Experiment 1. System B data was not as stable as we thought. The data extracted matched but wasn't in a format that could be used. We will have to transform the data before importing into the reporting system. This experiment FAILED.

- Experiment 2. System A data matched the reporting system data and the transformations performed on the data went better then expected. Transformation time was well below predicted times. This experiment was a SUCCESS.

- Experiment 3. Because Experiment 1 failed to transform good data to the reporting system, we are unable to figure out if any of the formulas are valid or efficient. This experiment FAILED.

The primary outcomes of the Immersive Experiences of the model are:

- Conduct or execute the planned experiments.
- Record the results of the experiments.
- Compare the experiment results against success criteria.

6. Critical Reflections

The team determined the root cause of the experiment failure. In this example, the reason for the failure was the tool used to extract the data wasn't configured correctly so the data extraction produced bad results.

We used the root cause analysis tool "5 Whys?" and we feel confident that changing the configuration might solve the problem. Although it's not possible in all cases to know the exact reason for an experiment failing, getting down to the most detailed explanation is essential as this information is used to create new experiments. Getting down to the root cause of the failure will ensure that the next round of experiments will be more successful.

Experiment 2 was successful for no critical reflections were needed on that experiment. We conducted critical reflections on those experiments that failed (experiments 1 and 3).

Experiment 3 was dependent on experiments 1 and 2 to succeed. Since we didn't run an experiment at all, no further critical reflection is needed.

Don't forget about the check list for this phase. Let's review the primary outcomes of the Critical Reflections phase:

- **Pattern Recognition** – look for patterns in the experiment results.

- **Potential Failure Causes** – identify any potential causes that caused the failure. Verify the experiments were carried out step by step in the same manner if comparisons are being made.

- **Acquire Feedback** – review your experiments and findings with others to determine if your approach and results are valid.

- **Baloney Detection Kit** – think skeptically about the results of the experiment.

- **Result Verification** – verify your results have valid data points.

7. Conceptualize Improvements

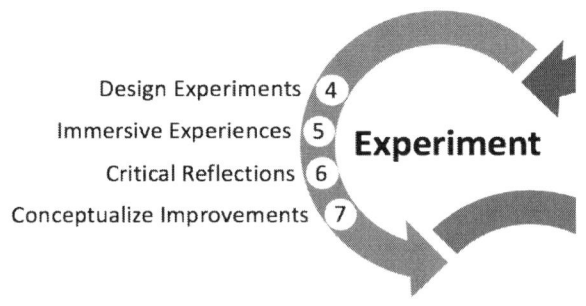

Based on our root cause analysis findings, we have determined the best course of action is to correct the configuration of the tool used in experiment 1. The team believes this will best solve the problem and should extract valid results. No changes to the success criteria are needed and the experiment doesn't need to change.

Here's quick reminder of the primary outcomes of the Conceptualize Improvements phase:

- Determine the root cause(s) – don't' guess the cause and make sure you are getting to the root of the problem to remove it entirely from making the experiment fail.
- Create potential experiments – during the conceptualize process you might discover the need to design a new experiment or change an experiment because you discovered that the experiment was not effective in helping your reach your success criteria.
- Prioritize potential experiments – Yes prioritize again. You may need to change the prioritization based on the outcomes. In this case because an experiment was a success, Experiment 3 moves up in priority and Experiment 1 remains at the top.
- Validate the success criteria – did the experiment meet the success criteria? Provide evidence that the success criteria was met.

8. Viability Analysis

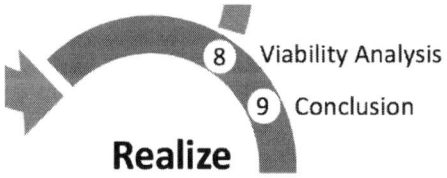

Viability analysis is about reviewing the outcome of all experiments and their relationship to each other. In thie case we decided to continue to experiment until the issues are resolved, but you could stop here if the experiments are no longer a priority by the organization or the organization has prioritized other experiments higher.

The primary outcomes of the Viability Analysis phase of the model are:

- Perform Viability Analysis
- Prepare for conclusion

9. Conclusion

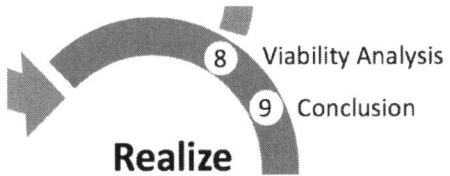

Looking at the model as a whole, what did we learn? How can we do it faster next time? What should never be done again?

Identify the lessons learned and action plans on how to take those lessons learned and improve. Don't just collect lessons learned and leave them sit on the shelf. Utilize them to figure out changes to make your processes work more effectively.

Chapter 8

Fail Safe Model

"A true friend is one who overlooks your failures and tolerates your success!"

- Doug Larson

Hollywood produced several Cold War thrillers about a nuclear crisis and how there was a Fail-Safe was in place to prevent a nuclear war. According to Merriam-Webster defines Fail-Safe as incorporating some feature for counteracting or reducing the effect of a failure. Our model doesn't focus on nuclear war rather on how to create systems and environments where individuals and organizations can softly land after a failure occurs. Learning from failure is the first step. The second step is planning for failure and failing safely.

The Fail Safe model is designed to help you and your organization fail safely. The components of the model below are a few of the many things you will need to consider when failing safely.

Let's dive into the most common areas where we can fail safely.

Micro Iterations

"More is not better. Better is better. You don't need a bigger house; you need a different floor plan. You don't need more stuff; you need stuff you'll actually use."

- Alex Steffen

Bigger isn't better. Keeping your experiments as small as possible makes performing those experiments easier. Experiments that are small can also be completed more quickly which in turn allows you to learn more rapidly. The principle here is to keep it simple and small.

Another advantage of utilizing micro iterations is that if the experiment is successful, it can be moved into a customer facing environment or be put into production more efficiently. This follows right along with Agile principles and approaches.

Developing a series of micro iterations requires a little more planning and alignment. One of the most significant challenges is not to lose sight of the overall picture. Experiments need to align to a central theme or goal. Check to make sure that as experiments are added or modified that they continue to remain within the context of the goal.

Keep a list of all the micro iterations and carefully track their status. Review the list and the experiments associated with them to ensure they keep within the context of the overall goal and are not deviating into unknown territory. Whereas documentation does take time, it will help guide your failure in the early phases of experimentation.

Track when you go larger than the micro iteration. Why? Was there a legitimate reason? Similar to Scrum methodology, when your sprint is dominated by one goal or result you are increasing the risk of failure that is harder to recover from.

Acceptance Criteria

We discussed the importance of acceptance criteria earlier. Always establish a baseline for what will be accepted as an outcome of the experiment. That baseline helps you safely fail because it clearly defines when a failure has occurred. If the expected outcome is to have 10 widgets produced, then experiment should produce 10 widgets. If the experiment doesn't create 10 widgets, that is the trigger to indicate the experiment failed. Without acceptance criteria set it's hard to know if the experiment was successful or a failure.

Make sure the acceptance criteria meet the overall goals and objectives of the organization. If the organization intends to produce 50% more widgets per day the experiment should have success criteria that measure the widget production contributing to that overall goal of 50%. If we provide 10 widgets per day currently, then the acceptance criteria and measurements must show we were able to achieve 15 widgets per day or 50% more (if that is the only measure – remember you can have multiple measures that add up to the 50% increase).

Acceptance criteria should be simple and easily measured. The more complicated a measure is to obtain the more likely that measurement will be inaccurate. Simple and easy to understand measurements allow for speed. If you can quickly gather measurements, it's probable to perform the experiment and obtain the measurement quickly. Interpretation of simple measurements and success criteria is also faster and less prone to error. Keep it simple in what you are measuring and how you obtain those measurements.

Is success criteria the same as acceptance criteria?

No. Acceptance Criteria is a term that is typically used in agile-based projects and is defined as the criteria for what the customer will minimally accept in the software. The acceptance criteria will relay what the system will do to support a user need. It relates to what is known as functional and non-functional requirements that are descriptors of what the solution will do. Examples of acceptance criteria: a first name, a password, a password that must have a numerical value, the password cannot be entered more than three times before being rejected.

Success criteria are referring to the results of the experiment. It forms a baseline to determine whether or not the experiment was a success or a failure. Success criteria is a measurable outcome for the experiment.

The subtle difference is the word "customer" and "experiment." Acceptance Criteria is focused on what the customer will accept in the final product. Success Criteria is focused on what the team will accept when the experiment is finished. Success criteria determines if the experiment has achieved its stated goal. Example: An experiment is being done to pay claims differently in the system by running the data about the claimant through a 3rd party system. The criteria for passing the experiment is that the claim will be paid 98.5% of the time using the new 3rd party system.

Know When to Fold 'Em

You have established success criteria, or a clearly measurable outcome, that you are hoping to achieve with the experiment. That's awesome. But how about the criteria to abandon the experiment? At what point are the results of the experiment so inconsistent and unrepeatable that you must stop the experiment? That is not an easy

If the experiment fails, how many times are you going to try to execute the experiment to see if you get the same result? Repeating an experiment to verify the results are consistent and not a fluke outcome will give you data with which to determine potential root causes for failure. Multiple experiment results will help you identify the root cause of the failure more quickly and accurately. Ideally running an experiment 2 - 3 times will give you a good set of results data, but that is not always the case for all experiments. The number of times you will repeat an experiment is something that would be established in the experiment design taking into consideration the amount of data the is needed to validate the result. As an example, for some experiments with human medical technology, the experiment would be carried out thousands of times while experiments involving web page design may be executed only 2 - 3 times.

Experimenting only once makes it very difficult to determine the root cause because you have only one set of data. With several sets of data, you will be able to start to see patterns. If the experiments all fail with the same results, finding the root cause is less complicated.

In an experiment with a software program designed to select the best shipping rates. The experiment is focusing on a specific ship to zip code. It will then automatically take the weight and dimensions of the shipment box into consideration when selecting the most cost-effective vendor for shipment. The results are not consistent. The success criteria states we will always get the cheapest rate. The success criteria were set based on data you previously downloaded from the shipping vendors. You know the expected result, but the results aren't matching it. We are getting different shipping vendors included in the results, but sometimes you are getting the expected result. Should we stop the experiment?

Try experimenting with different shipping box dimensions and weights. Set new expected results and try the experiment again. Did we get the same results? If we continued to get an answer that didn't match our anticipated results, does the success criteria need to change? What is the cause for the difference between what we are observing as the outcome and the expected outcome?

A quick investigation determined the experiment is flawed. The cause of the experiment failure can then be forwarded into the Critical Reflections part of the Fail Fast model.

Let's say after our quick investigation we immediately noticed why there was a difference. We noticed that we had connected to the production or "Live" version of the shipping data. The data we were relying on to make the calculation no longer matched what we designed for the experiment. In the "Live" version of the shipping data, shipping vendors change their rates frequently. Did the experiment fail? No, because the experiment itself was flawed by using the "Live" data from the shipping vendors.

Should you make the change to experiment to use the right data? It depends. If the change can be made to the experiment quickly and does not change the success criteria, then it seems reasonable. If the time and effort to make the change is significant or the success criteria needs to change, it is best to stop the experiment, note the failure and the potential reason for failure, then move the experiment into critical reflection where it can be analyzed further.

Setting the threshold that indicates how much effort can go into changing the experiment to get it to be successful is an important baseline. How many hours will you spend to fix the experiment? How many resources are you willing to dedicate to the resolution of the experiment design problems? A general rule of thumb answer is not to spend more than 2 – 4 hours to fix the design of the experiment. For resources, it is best to use as few as possible.

All the experiments you are performing can have the same threshold or a global threshold for when to stop attempting to fix the design. Can thresholds global? Probably not. As you design each experiment ask the question, "We set the threshold that after 3 hours of trying to fix an experiment design we will stop. Should that change for this specific experiment?"

Force Majeure Failure

Force Majeure is a Latin term that means superior force. It's a legal term that often refers to acts of God, weather-related incidents, or unavoidable accidents. You could even go as far as to say company mergers, acquisitions, reorganizations, and significant events happening in your industry are included in the Force Majeure definition. It is important to note the Force Majeure events are entirely out of your control, and you have no influence over them.

Force Majeure events will occur and affect your experiments. You need to be expecting them and understand how that event will change your experiment. Experiments may need to be abandoned, rescheduled, or even redesigned because of the large-scale event. When Force Majeure occurs, you will need to determine:

- Can the experiment continue?
- Does the experiment design still make sense in light of the Force Majeure event?
- Is the success criteria we established valid after the event happened?
- Do the environments for the experiment need to change?
- Can the experiment be put on hold and rescheduled?
- Has the vision and mission of the organization changed because of the Force Majeure event?
- Do we have the resources needed to continue the experiment?

These are a few items to consider when a Force Majeure event occurs.

Validate Environments

Safely failing is more achievable when the environments are in place for failure to occur. Environments can be anything from setting up a conference room to establishing a full-blown copy of the customer-facing system. Do you have all the systems, data, people, and processes in the current environment ready to fail quickly?

If your experiment is about a screen layout, do you have the ability to modify and update the screen components easily? Is making changes in this environment going to cause harm or difficulties for users of the screen? Are we in a place where we can try out new ideas without repercussions?

Where you fail is critical. As an example, let's say you are trying out a new kind of apple that was developed. You decide to have a group taste test the new apple to get their input, but you discover no one has any of the new apples for your taste test. They give you the old apples instead for the test. This doesn't create the environment you need to experiment by conducting a taste testing. You need to have the new apples available for taste testing.

If you are learning to downhill ski for the first time, your instructor doesn't take you to the black diamond professional slope. You start on the bunny hill and then progressively move up to more difficult levels (unless you wanted to show off, skip the bunny hill and break your leg - which is maybe not so much failing fast and failing safe).

This is the same with experiments. Start them in simple, safe environments and then scale them up to more complex environments. This will allow you to learn more about the outcomes of the experiments and gives you the opportunity to pass on scaling up. In the downhill skiing example, you may start on the bunny hill and then move to the intermediate hill and go no further depending on how well you did. By starting in a basic environment, it's easier to understand where the failures are occurring. When you start in more complex environments, finding those failure points are more difficult and can take significant time. Pick the environments that make the most sense for the experiment. Scale up the environments if needed to get more robust results and be cautious of moving too quickly into complex environments were failure results would be significantly more difficult to measure.

Does the environment support the experiment? If you execute the experiment is the environment large enough? Does it have all the data you need? Is the equipment used in the environment the same as the customer-facing environment? If the experiment environment and customer-facing environments are different, you may get significantly different outcomes to the experiments. If you understand the differences between the environments, you will be able to make a judgment call on whether the experiments will produce reasonable results to learn from.

Work with your team to make sure they have also reviewed the environment and elicit their feedback. This will help you avoid gotchas where environments don't operate as excepted.

Create a Safe Environment

Failing safely is about creating an environment where experimentation and learning do not cause harm or injury to yourself or your organization. An environment needs to be built in which experimentation can occur with wild abandon safely.

In the technology world, we use the term "Prototyping." Many prototyping environments are severely limited. If the environment is too confined or restrained for experimentation, then too few failures occur. It's harder to learn with so few failures. A better safe environment in one that this not restricted and open for experimentation. The best situation for prototyping is one that most closely the real-world environments.

Playing and changing everything in a production environment where your customer experiences your experiments tends to make your customers unhappy. Build an environment where you can play without consequence. You may have to start over from scratch and rebuild the environment after a wild day of experimentation. Plan on creating a way to restore your safe environment quickly after a mess has been made experimenting, so that future experimentation isn't slowed down.

Create other safe and soft-landing environments where you can bounce your ideas of others. Maybe your environment isn't about a physical space or system, but a room filled with flip charts and whiteboards.

Pulling together a group of colleagues to idea share, collaborate and innovate creates a safe environment if ground rules and expectations are set ahead of time. Set the expectations that experimentation and innovation is the goal. More ideas are better. We are not driving for perfection. It's like a brainstorming meeting on steroids. Encourage crazy ideas and try it out. There are no, and even the wildest crazy ideas are welcomed.

Another tactic is to experiment with screen or report design by having multiple variations mocked up. The key is to not just focus on one mockup but to have many mockups. A variety of mockups allows the group to "riff" off each other by taking elements of different mockups and combining them in new exciting ways.

One of our favorite tactics is user experience development and testing space. User experience folks will tell you it's preferred tactic to have users play with your interface (screen or report) and watch how they use it. Gather a group and invite them to play or experiment with the designs. The designers in the room are watching actual users interact with their design silently. The designers learn from watching the group play and experiment with the design. Designers then change the interface based on their observations. Rinse and repeat. One session is usually not enough. They key here is not to tell the user how to use the interface but to let them play and experiment freely in a safe environment.

Skills and Resources

Make sure you have the right people with the right skills to experiment. If the individual has little no experience with the equipment or process involved in the experiment, that might not be a good leader for that experiment. Try to find experienced individuals who understand and utilize the equipment used in the experiment on a daily basis. Learning is a big part of experimenting so having a less experienced individual shadow a more experienced user will provide the less qualified individual an opportunity to learn. By having both individuals that are experts and individuals that are inexperienced, you create an environment where broader learning can occur. Experienced users can sometimes be stuck in a rut and perform experiments in a specific manner because that is the way that equipment has always been operated. Less experienced individuals have more questions that can expose these "stuck in a rut" areas and challenge expert users to dig deeper into the fundamentals of the process or equipment.

If you are learning to downhill ski, you would most likely want to learn from an instructor that has good skiing and teaching skills. Resources should be knowledgeable enough to guide you out of situations of failure, researching why a failure occurred, and assisting in deciding if the experiment should be abandoned or not.

When learning to operate an ammonia sensor in a lab, we asked lots of questions on how it is operated. Some of those questions couldn't be answered even by the expert in operating the sensor, and we had to research them further. Experts can't remember every detail about their expertise and sometimes then need to research as well. It's also perfectly okay to ask lots of novice type questions because those questions allow a new perspective to be witnessed and understood. The reality is both the expert and the inexperienced individual wind up teaching each other. There are no bad questions. Smart people get smart because they ask stupid questions.

Setting Expectations

Setting expectations with your organization and team on failure will aid you failing more safely. Imagine if no one on your team understood why you are doing all these things that end up not working. They would feel pretty frustrated. This guy is just wasting our time!

Explain the purpose of failure and why it's essential. State it clearly that we are going to fail, and we are going to learn from that failure. This helps keep the team frustration levels lower. The communication on why we are experimenting, that these experiments are expected to fail, and to help us learn faster are messages that will need to go far and wide, so everyone understands why you are taking these steps.

Expectations can be set informally by email or talking about it in a meeting. In some situations, a more formal setting of expectations may be required by putting it in writing and getting signatures. Whatever method you choose depends on the experiment and the organization's ability to deal with experiments and failure.

Fail Safe Planning

Why bother planning? Just wing it, right? The message you will hear in business these days is that spending time on planning isn't valuable and just move forward. I'm sure you have heard the quote, "A plan shows how things are NOT going to happen." We have the unrealistic perspective that a plan can't change or can't change frequently. It may not be the plan itself that makes the experiment process move forward confidently.

"In preparing for battle I have always found that plans are useless, but planning is indispensable."

- Dwight Eisenhower, 34th President of the United States,

The real purpose of planning is to think through the plan from start to finish on how your vision or experiment will be accomplished. It helps build an image of the path you will be taking to complete the experiments or project. Planning enables you to minimize the unknown or uncertain parts of the journey to finishing the experiments or journey.

Planning is also an essential part of setting expectations. Planning help sets the expectations on the path that will be followed so that everyone has a general sense of the tasks and steps that are involved. The plan's purpose is to communicate the overall vision, mission, and critical steps that are needed so when these steps or tasks are occurring in the organization there is little conflict.

Nobody said you had to break out project management software. Some of the most significant achievements in human history were planned without a Gantt chart. Build a pyramid my followers, but first, let me see that Gantt chart. Entire civilizations arose without any charts.

Components of planning are:

- **Show the Big Picture** – how your experiments fit into the larger company structure and business mission statements and objectives

- **Break Down in Micro Iterations** – keep your experiments as small as possible, so they are not overwhelming to execute. The smaller, the better to ensure they are easier to manage and faster to complete. Smaller parts mean more attention is needed to ensure each smaller part is related to the larger experiment objective. Keep experiments aligned with the bigger picture

- **Plan Experiments** – understand how the experiment will be run, the desired outcomes, and the actual results.

- **Build Experiment Environments** – to ensure the experiment has an excellent place to research will all the equipment, materials and data needed for the experiment. Validate the environment reflects the real-world as much as possible.

- **Ensure the Right People are Involved** – who can perform the experiments and are experienced and understand what they are doing with the equipment, materials, and data. If individuals need training a specific type of equipment or material, be sure they get the appropriate training before the experiment starts.

- **Validate Experiment Data** – understand the data that will be involved in the experiment as an input and output. Validate the data you are using will allow the research to be conducted. Ensure the data used as an input into the experiment is reasonable and reality-based.

- **Define Measurements** – be clear on what success and failure mean. Define measurements that are easy and concise to determine if the experiment was a success or a failure. Ensure the measurements are meaningful in determining success or failure and don't use measurements that are not of any value.

- **Potential Points of Failure** – must be identified to see if the experiment can be adjusted to avoid these possible points of failure.

- **Pick a Soft Landing** – to ensure that if the experiment fails that you will land safely. Ensure that no harm will occur if the experiment fails. Don't jump out of a plane to test a new parachute, test it on the ground first.

- **Track Your Measurement** – make sure a system is in place to capture the data from the experiment and the results of the experiment.

Failing Safely

You shouldn't jump out of an airplane at 30,000 feet without a parachute. Setting yourself up for safely failing allows you to fail fast. Knowing that you have a soft place to land that will not have adverse effects on you or others will enable you to fail fast more easily. Failing safely takes way significant risks to you and your organization when a failure occurs.

There are many aspects around failure safely that we will explore. We first start out on how to convince your boss and others to try your idea because getting buy-in and support from others helps create a support system to allow you to fail safely. By building support for your idea, you form a network of leaders and colleagues that can support you in experimenting. We move on to talking points to discuss the value of experimentation to help you and the organization to learn quickly.

Setting expectations around failure before it happens failure sets the stage, so others understand that you are intentionally looking for failure and what you expect to learn from it. A large part of safely failing is handling conflict when a failure occurs. Addressing and managing conflict directly and positively creates a better environment to learn from failure.

More Safety Components

Every organization is different. There are many more ways in which to fail safely based on your organization type and industry. The Fail Safe model does not cover them all. Its purpose is to outline common areas that typically cause the most difficulties.

Take a moment and think about safety and failure in your experiments before you start experimenting. What things can you think of that are not listed here? Thoughtfully review other safety components and add them to your planning.

The Fail Safe model is meant to be changed and adapted to your organization. If you see other areas where you could fail more safely, why not add them to the model? Modify the model as needed to meet your personal or organization's needs.

Techniques to Accelerating Failure Safely

There are many techniques that can be used in experiments to accelerate failure and fail safely:

- **Sandboxing** – Do you have an environment where you can play with the system or solution before starting to build the solution? Be sure your prototyping environment is as close to production as possible. Allow participants to explore the new design on their own and observe their reactions. Try not to provide any training unless the participant gets stuck.

- **Prototyping** - Prototyping doesn't require any system environments. Prototyping can also be done collaboratively with a whiteboard. Example: You are modifying a report design. Participants walk through the design with the facilitator and make changes. You can use sticky notes as placeholders for report fields, so it's easier to move them all around the page. Design the new report collaboratively with participants. After prototyping is complete, building the prototype in a sandbox environment and performing the sandboxing techniques is a good fit.

- **Concept Piñata** – Beat up the concept until you get the candy out. Everyone gets to play the skeptic and challenges the concept. Once everyone has taken a turn bashing the concept, here's the twist. Pull the group into figuring out how to fix each item they identified as not working or things they didn't like. Take each complaint and flip it into something positive. Example: The screen won't work that way, but if you change the design to look like this then it will flow better. You said the color looked terrible, what color should it be?

- **Demolition Derby** – This is about taking several ideas or concepts and pitting them against each other. Determine a list of important capabilities, features, or things that are the most important for the product. Smash the ideas or designs against each other. One capability or feature at a time, compare each idea against each other. Try to see which concept performs it better. Some concepts will perform equally well for each specific capability or feature. Put a green check mark or mark next to the winning concept. Here's the twist. Everyone in the room working with this technique was focused on comparing the concepts together. What if we take the best from both concepts and put them into one excellent concept? The outcome of this doesn't need to be a selection of the current concepts being compared. The result can be an entirely new concept.

- **Conference Room Pilots** – Take a group through the experience of a particular process. This works well if a majority of the individuals in the room agree to the steps in the process before the conference room pilot begins. All the participants are going to play roles in the process. Determine what roles used in the process. Assign roles to everyone in the room. Walk through the process step by step with each role mock performing a step in the process. If there are reports or screen involved, try having them available in a prototyping environment where participants can interact with them in real-time. Otherwise, consider having screens and reports available in paper or hard-copy format. Gather their feedback as they walk through the process from start to finish.

- **Beta Version** – Develop a version for users to play with and get user feedback. Beta testing means the software is in an environment where they can interact with the software that is realistic or in the customer-facing environment. Let your users play with the concept and design using real-world data. Be sure you have an effective way to gather feedback.

Chapter 9

The Fear Model

"Even if you are on the right track, you'll get run over it if you just sit there."

- Will Rogers

Fear and failure go hand in hand, and for some it is paralyzing. Fear is a significant motivator in avoiding failure or how we react when a failure occurs. In some cases, we don't even recognize that we are acting out of fear instead of logical reason. Fear can significantly limit your ability to Experientially Learn because instead of learning from the failure we fear it and then avoid it. In other cases, we don't understand that full nature of what we are afraid of, but instead, make a snap judgment.

The best approach to managing fear is to define the boundaries and context of the fear fully. Understanding the fear from all the angles is an important step. It is difficult to impossible to try to manage fear when you don't fully understand what triggered the fear.

Our human nature is to respond to threats or fears in a split second without conscious thought. The faster you respond to danger the more likely you are to survive. While that is that case in the wild or life-threatening situation, it might not be the best approach within an organization.

Taking a step back and getting the lay of the land is an important step. Take a breath and give yourself a chance to observe the situation more thoughtfully. Making decisions without having a good picture of what triggered the fear sends you into a chaos of decisions that could wind up making the fear even more disastrous. There are of course knee-jerk reactions that are in your best interest. Dodging a speeding car coming at you down the sidewalk doesn't require an in-depth thought process of why. In organizations, most decisions are not as life-threatening, and you can take a step back and understand it further.

Your colleagues and leadership will pressure you to act without stepping back. Real leaders make quick decisions, right? Good leaders will step back and assess the situation before making any far-reaching choices to ensure further action will not add to the impact of the situation on the organization. "This seems like a knee-jerk reaction. Can we step back for a few minutes and understand what just happened more fully?" is a helpful statement to use in these situations.

The initial reaction to fear can be filled with strong emotions. Take a step back and constructively express those emotions. "I understand your anger and would be angry too," creates a better relationship that confrontation of facts and figures. Let the emotion come out but move quickly to a place of comfortable fact-based decision making.

The Fear Model

The Fear Model is an approach to handling both individual and organizational fear when dealing with failure. The approach of this model is to provide the reader with an easy way to address common fears.

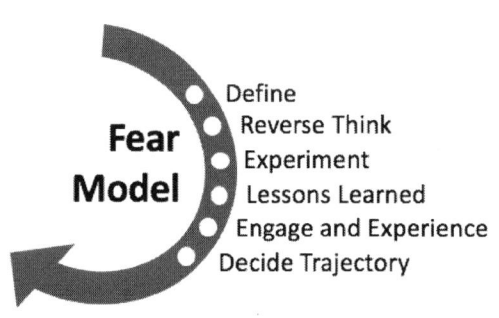

This model is designed to "Lather. Rinse. Repeat." Going through this model multiple times to address fear and resolve that fear is common. Fears can be complex with many different layers. Break larger more complex fears into smaller components that will be easier to handle when going through the model. Take each smaller part of the overall fear through the model one at a time. Taking large complex fears as a whole and putting them through this model can be counterproductive.

Let's define each step in the Fear model in more detail.

1. Define and Identify
A situation has occurred, and we have experienced fear. Fear is the start of the model. We notice ourselves wanting to either avoid it or feel strong emotions towards it like anger. An essential part of this step is immediately recognizing fear is happening. Not an easy thing to do when you are in the heat of the moment, and it does take some practice.

2. Observe and Reverse Think
In this part of the model, we need to step back and observe the situation from a broader perspective. Avoid being forced to act. Get to the root cause of the situation to understand why the fear is occurring. Look for gaps in your thinking that you are filling by making assumptions. Try to focus on removing any assumptions. What is the actual context and scope of the situation? Why are you experiencing this fear?

3. Experiment Safely
Devise and execute experiments that will allow you to interact with the fear in a safe and meaningful way based on your understanding. Look for solutions to the root cause of the fear. See experiments as a way to further understand the situation more comprehensively and

devise possible solutions. Give yourself the permission to fail and understand that failure at these experiments is how you learn.

4. Lessons Learned

Thoughtfully reflect on the experiments conducted to understand their outcomes. Create a plan based on the experimentations that you believe will solve or make the situation less fearful.

5. Engage and Experience

Execute the plan to resolve the situation or fear in a real-world environment. Failure can and will happen in the real-world when other factors present themselves that were not considered during the experiments or expected to occur. Realize that this failure is an opportunity to learn more about the situation.

6. Decide Your Trajectory

Don't beat yourself up for having failed. Engage those individuals that helped you resolve the fear to see their point of view on setting your trajectory. Your trajectory is what you will do next with the fear after experimenting with it. Decide whether the outcome is continuing to experiment with the fear or entirely abandon all experiments on the fear.

Celebrate the success! Find a way to congratulate and be grateful for those that assisted you in resolving the situation.

After reading through the model, we are sure you got the impression that this model looks almost identical to the Fail Fast model. There is some overlap between the models. We designed the Fear Model around advice from behavior experts who understand the fear and how to overcome fear. The Fail Fast model is designed to help you work with a discovery, experience failure, and learn from that failure.

Using the Fear Model – The Engagement Party

The names of this story have been changed to protect the innocent. Our main character in this story is "Sullivan." Sullivan related this story to us about an engagement party and his fears.

"A friend and I were out having dinner when she announced her engagement. After an eyeful of a huge sparkly rock on her finger, I gave her some hearty congratulations and told her if she needed anything to let me know. A twinkle appeared in her eye, and she asked me to host the engagement party. My reaction was to fall back and offer to buy some food and wine. My friend tells me that I did give some pretty good parties in the past and it would help her out to have someone take all the engagement party stuff to off her plate. She wanted an elegant soiree with adorable little appetizers and sparkling champagne. After some negotiation, I did finally agree to host the party."

Let's take this story and use it to walk through the fear model.

1. Define and Identify
Sullivan is afraid of hosting parties and cooking for others. He's hosted parties before with pizza and beer, but that it a far cry from an elegant soiree. Sullivan had to step back and figure out why he was afraid of throwing a more formal party. He realized he didn't fully

understand all the formal bits of a party. Calling the pizza shop down the road wouldn't work for the food. He needed to figure out how to make those adorable, elegant appetizer things. Sullivan's fear was the unknown.

2. Observe and Reverse Think

Sullivan can burn water, and his best dish is a microwaved hot dog. He needed to learn some good appetizer recipes for the elegant soiree. "What does an elegant soiree look like?" Sullivan wondered. He researched a bit and figured out more about these elegant soiree and engagement parties. They were looking like a lot more work than hot dogs in the microwave. After finishing his research, he took an inventory of his appetizer recipes. Sadly, lacking in that department, he searched for recipes online that were easy. Sullivan isn't a professional chef. He needed things that looked and tasted good but were easy to make. He figured out his boundaries of what he could accomplish with his skills. Sullivan reviewed his thinking to make sure he didn't have any assumptions about his skills.

3. Experiment Safely

After thinking it through a little further, Sullivan figured it was probably a good idea to test out some of the recipes he acquired. Sullivan also thought about a test run before the actual event happened. All that project management training paid off. He decided to hold a few dinners before the engagement party, test everything out, and ask for feedback from his guests. He'll make mistakes but was determined to have fun and make it work. Several of his first attempts were a disaster, but after a while, he was getting into the hang of it.

4. Lessons Learned

After a few failed attempts and experiments, He learned how to organize himself better in the kitchen and to have things prepared in advance, so he will able to be a better host. He shouldn't be in the kitchen putting out fires - literally putting out flames. Nothing says elegant like the smoke detector going off. He wanted to appear calm and in control. He took his lessons learned and formulated a plan on how to pull it off. He couldn't do everything himself, so he enlisted the help of a few friends. Ok, let's hold that engagement party!

5. Engage and Experience

It was show time! The engagement party was a success by the feedback of everyone that attended even though he did have a few failures in the kitchen. Practice and experimentation ahead of time prepared Sullivan for what he would need to do in the event of a failure. Sullivan thanked his friends and was very grateful for their work. Stepping back again, he thought over his accomplishments and lessons learned from having this engagement party.

6. Decide Your Trajectory

The engagement party went pretty well. Sullivan had less fear of hosting a formal party. He might do it again in the future! Even if it didn't go so well it was good to recognize a potential future direction he could take.

In addition to the model, there are other factors to consider when dealing with fear. Let's discuss some of those factors at a high-level.

The Ladder of Inference

We can make some interesting conclusions when confronted with a fearful situation. Chris Argyris[60] was an American business theorist and Professor Emeritus at Harvard Business School who is credited with the Ladder of Inference model. This model explains the steps we go through in our minds when presented with new situations -- how we make inferences. An inference is a conclusion reached on the basis of evidence and reasoning.

Here's an example of how the Ladder of Inference works. You are driving down a side street slowly in heavy traffic. Suddenly a large pickup truck appears in your rearview mirror weaving through traffic at high speed. You worry he will hit your car. He zips past you and nearly clips off your bumper attempting to pass you. You see him yelling something out the window of his truck as he passes by. This man is not observing the speed limit or safe driving protocol. You think you should shout it out loud that this man is an idiot. You noticed the truck was older and rusted so therefore it must belong to some deadbeat who is probably angry and driving drunk. You honk your horn loudly and yell at the man in the truck. Wow. How did we get there?

---

[60] Wikipedia, en.wikipedia.org/w/index.php?title=Chris_Argyris&oldid=837307208, Chris Argyris, Various Contributors, April 2018

Understanding the Ladder of Inference

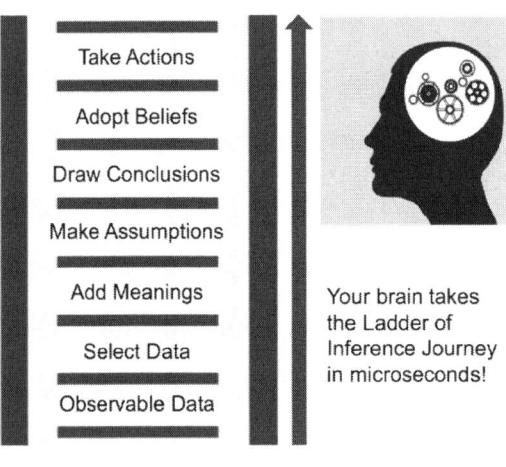

Your brain takes the Ladder of Inference Journey in microseconds!

Observable Data

Imagine that everything you see is a 30-second YouTube video. In that 30 seconds, you only see what you want to see. Your brain captures all of it, but it stores it in places that we mostly ignore. You see what you want to see.

Select Data

Now you select the data that you saw – the parts that make sense to you or register with you.

Add Meanings

Here is where it gets complicated. You add what we call "personal baggage." You are adding meaning based on your experiences, culture, race, religion, orientation, and value systems.

Make Assumptions

Based on the meanings you have added to your selected data, you make assumptions about what is happening. We all know what happens when we assume. You make an…

Draw Conclusions

Once these things are in place the conclusion(s) has been drawn.

Adopt Beliefs

With the drawn conclusion, it fits into a world-view of how things must be, and in your brain there are no other options at this moment.

Take Action

Take action. Wrong, right or indifferent some action is taken.

With that understanding of what the Ladder of Inference is and does, let's go back to our pickup truck example.

Inference Ladder Example

An observation is made. You observe something without bias as a witness or bystander. You notice a truck approaching you quickly and weaving in and out of traffic in your rear-view mirror. It is just data at this point.

You select data from what you observe. The rear-view mirror, older rusted truck, weaving in and out of traffic and traveling at high speeds on a side road. The key here is "selected" data and not complete data. You saw what you wanted to see or only could see.

You deduce meaning to the selected observations. Based on the data you observe, you draw from experience. People who are unsafe drivers drive an old rusted out truck. You've experienced this before. That would never happen to you with your upbringing.

Gaps are filled with assumptions. The selected data we choose to observe didn't create a full picture in our minds. There were gaps, so we filled them in with assumptions from our experiences. You didn't know why the truck was driving fast, so you fill the gap with an assumption the driver was angry, irresponsible, reckless, or dangerous.

Conclusions are drawn. You conclude the situation by creating emotional responses. In this case anger. This reckless driver is going to endanger lives. You are likely to get hit!

Beliefs are adopted. This person is a menace to society and a deadbeat. Something must be done.

A response or action must be taken. Therefore because of this conclusion and emotion, you must take some action even if it is inaction. You honk my horn loudly and swear at the other driver. You call 911.

Is that what really happened? This might have been the reality. The man was rushing to the hospital that was a block away. He had sharp pains in his chest and was racing to the hospital emergency room. He most certainly did not make the best choice in responding to his chest pain by driving at unsafe speeds in that he could have died while still speeding down the road. What was he yelling out his window? He was yelling an apology out the window of his truck to other drivers for cutting them off.

If you get a chance, check out the story of Joshua Bell online. It is a very well documented story (that we have used in other books) of the Ladder of Inference at work. There is also some great video of the experiment.

Walking through the ladder of inference is done in microseconds. It allows us to think and react quickly. In organizational settings where life and death decisions are not the norm, this type of thinking can lead to decisions that are based on assumptions and conclusions that didn't consider the entire situation. By stepping back and being aware of how the organization or individual is walking through the inference ladder is an essential step in controlling fear.

The other advantage of stepping back is understanding what your inner voice is saying to you. Is it the negative voice that says you can't? Is it feeding your anger or fear? This inner voice can be compelling. Think about what your inner voice is telling you, so you can reason it out to be valid or not.

What to overcome the fear? Smile, laugh and exaggerate it. You might not be able to do this in front of others who are having trouble coming to grips with their fear, but it is a way to deal with your internal fears. If you laugh at the most horrible outcome, then you are lessening your anxiety. Laughing causes positive endorphins to be released even when you are faking that laughter. This fake smile and laughter assists in lowering your stress levels. It's not about cutting down other people, but instead, it is cutting down the issue at hand, and you take away a lot of its power over you. It doesn't mean the issue is less real or less important because you cut it down in your mind. It's a trick for dealing with a difficult situation more effectively.

At the time of writing this chapter (literally), Bob needed to laugh off the fear. What happened? Bob got a new laptop and iTunes somehow lost all of his music data (a service he had been paying for). Bob had to download all of his music from a backup drive (less 2 years' worth of music purchases that can't be recovered) onto his computer. Now before you lament that situation or rail on iTunes, that is not the issue. The issue was Bob had cellular data turned on his phone. Music was downloading to his phone, and somehow the Wi-Fi went out during the night. Cellular kicked in. Bob woke up to dozens of texts about his going over his data plan and AT&T adding another 1 gigabyte for $15. Damage has been assessed, and this is going to cost Bob $2,000.00. Ladder of inference, fear, everything has kicked in. Imagine you have to tell your spouse that you just cost them $2,000.00 for downloading music on your phone that you did not want on your phone to begin with. This is a bigger failure than Bob would like. It is a valuable lesson to do many more experiments with iTunes, AT&T, the phone. All Bob can do right now is to laugh it off. A lot. Then maybe cry a little and laugh some more!

As leaders, we have asked the questions, "Did someone die? Could someone die?" when the issues presented themselves. We were always amazed at how stressed out someone could get at a database field not being populated correctly and heading straight into panic mode. Panic gets you nowhere. The same is true of intense emotions. On more than one occasion we have noticed that because we are calm and collected, the entire team (no matter how hard they tried) couldn't help but to calm down with us. The same is true of when we were in contributor roles or members of the team. It's a compliment to be called even headed. Others always valued that because it allowed us to think more clearly and objectively to create better conclusion and solutions to the situation.

Always tackle fear immediately. Whether it is your own or your organizations fear. Avoiding fear will only make it grow and get more out of control. When fear gets too big, it takes on a life of its own and can't be reasoned with. Addressing fear doesn't mean barreling into the situation like a bull in a china shop. Tackle fear directly by wanting to understand it in a more meaningful way.

External Influences

How we react and deal with fear is also based on external influences. What we are told affects the way we think about ourselves or our organization and defines how we approach and handle failure.

Understanding external influence is discussed in Carol Dweck's[61] research study about external influence on achievement and in the

---

[61] Wikipedia, en.wikipedia.org/w/index.php?title=Carol_Dweck&oldid=859394427, Carol Dweck, Various Contributors, September 2018

book "NurtureShock: New Thinking About Children" by Po Bronson[62] and Ashley Merryman[63].

In her experiment, Carol Dweck had four hundred fifth-grade students complete a test composed of easy-to-solve problems to look at external influences on achievement. Students were put into two groups. The first group was always praised for their intelligence. The second group was always praised for their effort. Both groups were given a simple test first and praised accordingly. The intelligent group was praised for their intelligence. The effort group praised for their effort.

Next, both groups were given a choice between taking two tests. One test would be as easy as the one they just received. The other test would be more difficult, but they would learn more from it. The majority of students praised for their intelligence chose the simple test. The opposite was true for the students that were praised for their effort. The students who were praised for their effort not only performed better on the harder test but also had a more positive reaction to their performance on the harder test.

Next, the harder test was given to both groups. The intelligence praised group didn't perform as well as expected and had a more negative reaction to performing worse on the harder test. The effort group performed better on the harder test for the second time because they learned from taking the previous difficult test and

---

[62] WikiPedia, en.wikipedia.org/w/index.php?title=Po_Bronson&oldid=844632489 , Po Bronson, Various Contributors, June 2018

[63] Ashley Marryman, www.ashleymerryman.com/, Ashley Merryman, August 2018

applied that knowledge to the new harder test that was presented to them.

This experiment shows how we are influenced by our leaders and others when dealing with failure or risk. Be aware that the influences of a leader can assist or deter you in helping to harness failure to learn.

Organizational Influences

If the organization rewards its members for being risk takers, the individuals in that organization will embrace a risk-taking attitude. Fear will still be there but not as much as the organizations that frown on risk-taking. Rewarding risk-averse behavior will create a culture of risk avoidance. The attitudes you reward become the prominent attitudes in the organization's culture. Do you want your people to deal with more or less fear in your organization?

Be careful with organizations that state one thing but do another in practice. Many companies say they are ok with risk or failure, but when confronted with that risk or failure they will avoid it entirely. Understand that organizations have a "Talk-the-Talk." These are the things that they say out loud, write down and post on the walls, but never tangibly deliver. It's better to understand organizations may not always embrace failure by observing their actions, not their words.

Many organizations say they want to fail fast, but this can mean they are a learning organization that is still constrained by funding, market and time, so the failure that is occurring is happening in tiny increments. A macro failure cannot be tolerated, so failures are kept as small as possible. Small failures still provide opportunities to learn although the magnitude of the learning is smaller.

Large-scale failures are generally not a good plan. If the failure is too massive, it will adversely impact the organization's reputation, customers, and product. Keep in mind this will depend on the industry and what constitutes large-scale failure. Having worked in Fortune 50 companies for 20 years, million-dollar failures happen all the time without any repercussions.

Organizations need to set a formal threshold or tolerance of failure in each of their environments. Understanding these tolerances will help deal with the associated fear of failing. That threshold needs to be established for the public product environment but also for the R&D and pre-production or pre-release environments in the product testing lab. Organizations that have a zero tolerance for failure in the public version of their product will set high standards for their product in the testing labs or test environments.

Different environments will have different tolerance levels for failure which means they will all have different attitudes towards fear.

In the Research and Development environment in an organization, the tolerance for failure is typically quite high. High tolerance means a tremendous amount of failure can occur. In this R&D environment, a significant amount of learning is achieved because the failure rate is quite high.

In pre-production or pre-release environments, the tolerance is lower than the R&D environment. Typically, this is more of a medium tolerance for failure. Medium tolerance means some failure can occur, but it is limited in scope and size. Large failure in this environment isn't tolerated as this environment is used to test the product before its public release.

In the production environments, most organizations have a small to zero tolerance for failure. The tolerance is low due to the fact the product is now public facing, and the organization wishes to protect its reputation.

An organization can have multiple layers of environments for their products. We have worked with companies that have as many as ten different environments. Each environment had a failure tolerance established that was well defined. Some people laughed in the face of fear and others cried.

If the failure threshold is crossed, the product is sent back to the previous environment for further experimentation and failure. Sending the product back to an earlier environment allows the environment to remain stable enough to ready the product for the public. Well established tolerances at each level ensure a smooth transition between the environments.

All of these different levels of tolerances create the external influences and language used for the organization to reward – or punish – individuals within the organization. Keep in mind that the organization's leadership sets the tolerance levels, and those tolerance levels may not be well established thus leading to stressful and fearful situations when a failure does occur.

Conformity Drives Organizational Fear

Solomon Asch[64] was a pioneer in social psychology and is best known for his conformity experiments. The most significant outcome of his studies was that peer pressure even in adults could be influential on an individual's decisions.

His conformity experiment was conducted using 123 male participants. The male participants were told that they would be part of an experiment in visual judgment. Each subject was put into a group with 5 to 7 insiders that knew the real aim of the test but whose identities were withheld from the participants.

The group was then shown a card with a line on it. The next card shown to the group was a card with three lines on it that were similar in length. Each line on the card was labeled A, B, and C. The group was then asked to say A, B, or C out loud for the line they thought matched the first card. The real participant always went last in answering.

For the first couple of rounds of the study, the insiders always answered correctly. The real participant always answered correctly as well and felt comfortable with their choice.

In the next few rounds, some of the insiders intentionally answered incorrectly while other insiders answered correctly. The real participant could go along with what they clearly could see as the

---

[64] Wikipedia, en.wikipedia.org/w/index.php?title=Solomon_Asch&oldid=856669477, Solomon Asch, Various Contributors, August 2018

correct answer or merely side with the majority. The real participant was to free to agree or disagree with the group's answer.

In the last final rounds, all of the insiders answered incorrectly every time. The real participant was still to free to agree or disagree with the group's answer.

However, in the final round, all the insiders responded with a wrong answer in 12 out of the 18 trials. The trials in which the insiders answered incorrectly were the "critical trials." The participant could go with his answer or go along with the majority and ignore the obvious fact. The aim was to see whether the real participant would change his response and respond the same way as everyone in the room or stick with what his eyes plainly told him.

The study found that some 20% of study participants withstood the peer pressure to vote incorrectly. About 1% of real participants succumbed to the majority regardless of their answer. While the remaining 79% waffled on their answers believing the group answer to be more correct than what their own eyes indicated. Asch suggested doubt was created in the minds of the real participants upon hearing the wrong answer. Real participants reported seeing the line matching the incorrect answer, but began to doubt what they were seeing. The study revealed that when 1 to 3 of the insiders in the group answered incorrectly, it influenced the real participants significantly. However, that effectiveness did not increase when 3 or more insiders responded to the question incorrectly.

Conformity in an organization can influence decision making and perception of the failure. To learn from failure effectively, an organization must ensure individuals and its leadership are not influencing each other. Leaders need to embrace differing points of view and individuals that bring up uncomfortable issues. By shutting down individuals, the organization isn't able to get a full perspective. Be aware of this influence taking place and challenge it appropriately by keeping the group away from opinions of why the failure occurred to more concrete data and critical thinking. Always be respectful of other's opinions but ask that those opinions be backed up with reasonable evidence to support the opinion.

Ask yourself, "Am I just going along with everyone to not cause issues?" If you answer "Yes" to the question, you are conforming to the group. Many bad decisions occur because one person didn't speak up. Assert your opinion and ideas where needed being careful not to dominate the conversation or be disrespectful to others' viewpoints.

Tidying Up Your Fear

"You gain strength, courage and confidence by every experience in which you really stop to look fear in the face. You are able to say to yourself, "I have lived through this horror. I can take the next thing that comes along."

- Eleanore Roosevelt

Remember, if you want to deal with your fear, practice the steps in the fear model:
1. Define and Identify
2. Observe and Reverse Think
3. Experiment Safely
4. Lessons Learned
5. Engage and Experience
6. Decide Your Trajectory

It takes time to get used to dealing with your fear on a regular basis. You must develop new habits. Take time out every day and reflect on becoming a new fearless you.

Chapter 10
Agile & Failure

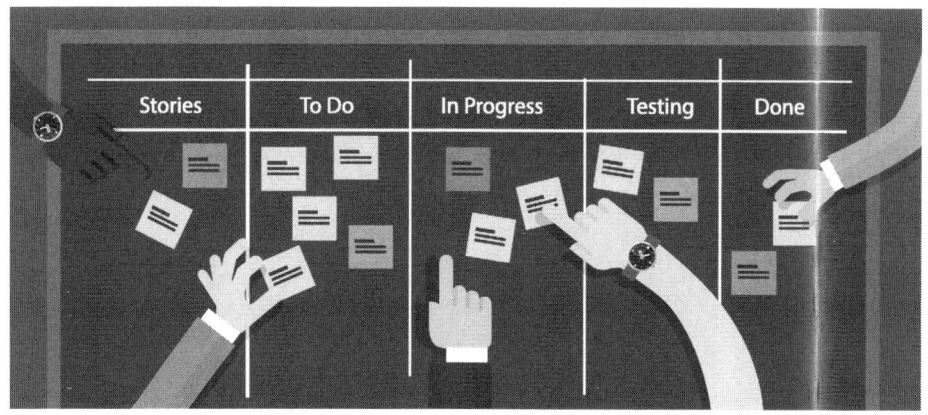

"Success today requires the agility and drive to constantly rethink, reinvigorate, react, and reinvent."

- Bill Gates

Agile & Failure: Why You Won't Stop There

Agile, Agile, Agile! Marsha, Marsha, Marsha! Everyone is talking about or doing Agile (just like everything was about that character Marsha in "The Brady Bunch" television show). What is the big deal? Does it impact everything we have been reading here or has this book just been a veiled attempt to sell Agile?

Technology and technical folks will be able to relate to this chapter a little more strongly than individuals with careers outside of the technical and technology realms. Our focus is to bring the technical and non-technical folks together into a shared understanding of Agile and the Fail Fast model. You don't have to be technical to be Agile and Fail Fast and Safe.

Let's start with Agile. Some people see Agile as a way of failing fast, but there are others who see Agile as something completely different. Let's explore Agile and how it fits into the Fail Fast model and how it relates to both you and your organization.

When you hear the word "Agile" what comes to mind? If you could explain what it means to you in just one word what would that be? Now close the book, put it down, look away and say that word before you read on.

Here are some words that our clients often say when we ask, "What does the word Agile mean to you in just one word?"

Fast	Easy	Micromanagement
Savings	Process	Buzz-word
Methodology	Principles	Sucks
Deadline-less	Scrum	Cheaper
Explore	Autonomous	Agility
Dynamic	More-productive	No documentation
Empowered	Experiment	Collaboration
Iterative	Flexible	Fun

What was your Agile descriptor? It might have been in the list above or something entirely different. Some of the words to describe Agile are more myths than reality while some of these words are more representative of Agile's reputation. There are often misunderstandings between the definition of Agile and how Agile is performed.

Let's start with the definition. The dictionary states Agile is:

Agile

ag·ile [ˈajəl]

adjective

1. able to move quickly and easily

Is Agile quick and easy? It can be, but not for most companies. Many companies are struggling with Agile concepts and delivery. They are finding Agile not to be quick or easy. This is not a fault of Agile but a fault of a company's implementation of it.

The Agile Alliance states that Agile is "The ability to create and respond to change to succeed in an uncertain and turbulent environment." That very definition reinforces failing fast and failing safe. Other definitions might say it is an iterative software development approach that builds software incrementally versus delivering everything at the end of the project. You also have additional concepts like an Agile mindset versus Agile methodology. It is no wonder why people struggle when they start going Agile. So, what is Agile?

Foremost, Agile is a set of principles to live by when developing software. It is not a methodology. It is a mindset that you have that increases the chances of delivering successful products. Once you understand Agile principles, you can then determine which Agile Principles can work in your organization. Then you choose the Agile methodology that uses the principles, which is the process that best fits your organization's needs. To figure out how this all fits into Fail Fast and Fail Safe, we are going to start at the beginning with Agile principles.

In the 1990's, several software development methodologies were gaining traction as a better way of delivering business value based on the frequency of delivery and self-organized teams as an alternative to the documentation heavy software development process that most companies were doing (Waterfall). A group of 17 software development practitioners got together to share their ideas, and from that, they produced the Agile Manifesto.

Many people hear the word "manifesto" and immediately it produces a negative response. We understand why. The dictionary states a manifesto is a public declaration of policy and aims, especially one issued before an election by a political party or candidate. People immediately think of Karl Marx, Adolph Hitler, The Ten Commandments, The Declaration of Independence, and more. The term "manifesto" is unappealing for many reasons across history (even in today's challenging political climate). When we typically describe Agile, we try to stay away from the manifesto term and instead focus on the set of 12 Agile principles the group created. Much like this book, don't think of it as a manifesto, think of it as inputs into potential success.

We believe in the concepts of Agile and the models outlined in our Fail Fast model, Fail Safe model and Fear Model. Don't take them literally. These are all models that are tools to reduce risk and increase success.

Quick side note, there is an urban legend of an interview with Winston Royce (credited with creating Waterfall), and we cannot verify this urban legend, but it does make sense. He was quoted "I didn't think anyone would take it literally." We tend to believe this legend because in his paper "Managing the Development of Large Software Systems" he states after he introduces the pattern that became Waterfall. "I believe in this concept, but the implementation described above is risky and invites failure." Winston Royce said. If you dig into his writings, you can see he does promote many Agile-like concepts.

"We can't solve problems by using the same kind of thinking we used when we created them."

- Albert Einstein

18 Flavors of Agile and Going Strong

When we are talking about Agile, we are referring to the 18 different methodologies aligned with Agile or that have been identified as being Agile. They are:

1. ASD (Adaptive software development)
2. Agile modeling
3. AUP (Agile Unified Process)
4. BADM (Business analyst designer method)
5. Crystal Clear Methods
6. DAD (Disciplined agile delivery)
7. DSDM (Dynamic systems development method)
8. XP (Extreme programming)
9. FDD (Feature-driven development)
10. Lean software development
11. LESS (Large Scale Scrum)
12. Kanban (development)
13. Nexus (Scaled Scrum)
14. RAD (Rapid application development)
15. SAFE (Scaled Agile Framework)
16. SCRAP (Scrum/XP hybrid)
17. Scrum
18. Scrumban (Scrum/Kanban)

Many books will describe in detail how those methodologies work, and we will leave it to them to help you on your Agile methodology journey. However, the point of elaborating this list of methodologies is to show them all and demonstrate there are several methodologies that strive to overcome complexity by failing faster. By failing more quickly, they are learning faster!

Rather than failing after a year of development, Agile helps us to fail faster. Failing faster allows us to quickly learn from that failure, critically reflect on it and take action to conceptualize improvements.

Agile Principles

Here are the Agile principles the seventeen software development practitioners came up with:

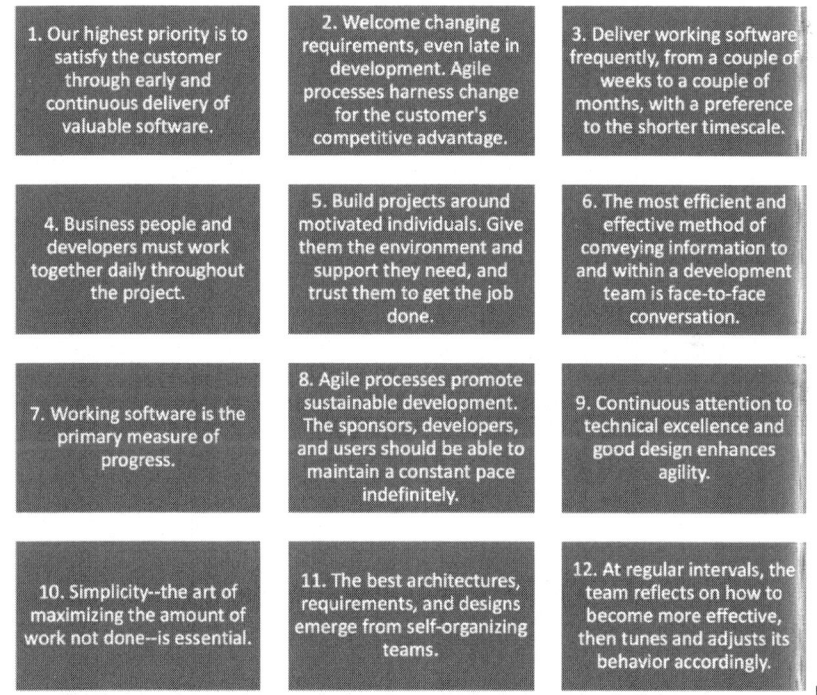

1. Our highest priority is to satisfy the customer through early and continuous delivery of valuable software.

2. Welcome changing requirements, even late in development. Agile processes harness change for the customer's competitive advantage.

3. Deliver working software frequently, from a couple of weeks to a couple of months, with a preference to the shorter timescale.

4. Business people and developers must work together daily throughout the project.

5. Build projects around motivated individuals. Give them the environment and support they need, and trust them to get the job done.

6. The most efficient and effective method of conveying information to and within a development team is face-to-face conversation.

7. Working software is the primary measure of progress.

8. Agile processes promote sustainable development. The sponsors, developers, and users should be able to maintain a constant pace indefinitely.

9. Continuous attention to technical excellence and good design enhances agility.

10. Simplicity--the art of maximizing the amount of work not done--is essential.

11. The best architectures, requirements, and designs emerge from self-organizing teams.

12. At regular intervals, the team reflects on how to become more effective, then tunes and adjusts its behavior accordingly.

Look at the Agile principles above. Which of these Agile principles are the most difficult for you or your organization to implement?

When we show these to clients and students the typical responses we get are:

- All of them!

- We are so compliant that will never happen.

- We can keep changing requirements. We have to nail it down and move forward.

- How do you develop working software when everything we are doing is infrastructure that takes months to build? That does not compute.

- We are global. The business would never give up someone to sit with IT.

- They don't trust us. They change their minds all the time. They don't let us finish what we start.

- As I said, we are global. Face to face does not work. They do not give us the tools. Offshoring is a reality. We get an hour, each day to talk and that is not enough.

- It is not the primary measure at our company. It never will be.

- Um, people quit. People take vacations. That sounds a lot like forced labor to me.

- How? We don't have time to sit and review and be technically excellent. They are throwing so much at as constantly that we can barely breathe.

- What does that even mean?

- So, the team decides right? We are such a top-heavy organization that our voices are constantly discounted.

- You are talking about lessons learned. No one does anything with them other than store them in a drawer or a folder on the network.

With all of this negativity, does this mean Agile is terrible for you? Of course not. What we have to remember is that Agile is a set of principles that when you use them, you have a higher chance of success, but often Agile is met with disdain, skepticism, and doubt which is understandable given the bad habits that have formed in most organizations. When the habits change, the success with Agile will start to come.

The comments above are coming from a mindset that doesn't understand failure and how to harness that failure for learning - yet. As the organization starts to understand the concepts of the Fail Fast model, it will find itself embracing the Agile mindset more effectively.

Embrace Failure Because Learning Can Never Stop

A significant part of Agile or any other methodology is failing and learning from that failure. When you start a project, do you have a detailed design just ready to go? In most cases, we need to take a step back and learn about the product we are creating or modifying. We continue to learn by going through the design process and elicitation of capabilities and requirements for the new product. Throughout the entire project, we will try something, fail, and learn from that failure.

Have you been to a retrospective lately? These are designed to help the team understand what was learned. Even the daily scrum meetings (standups) themselves can be used to support the team in understanding what we have learned (please note we are saying support and not how to solve – solving is not what your standups are for). Do we take what we have learned and incorporate it into our process, design, or product? In most cases, we do not. We typically learn something valuable and don't do anything about it. We are missing the opportunity to move forward with a stronger understanding that can propel the project forward faster and more successfully.

Look again at the agile principles above. How does failure and learning from failure work with those principles?

Let's take "Welcoming changing requirements, even late in development." as an example. This principle speaks to flexibility and adaptability. What happens when things change rapidly? We get frustrated because things seem out of control and we don't understand that change can drive failure. Subsequently, this failure allows us to learn.

Let's apply this to the steps in the Fail Fast model. When we fail, we were unable to meet the expectations of the customer. We can then take that failure and learn from it in the Critical Reflection phase. Because we gain a better understanding of the failure, we can discover solutions in the conceptualize improvements phase. Then we build and deploy the solution in the Design Experiments phase. After the solution is deployed, we go back to the Immersive Experiences phase to see if the solution implemented will meet our customer's expectations. In the Immersive Experience phase, both you and the customer interact with the solution in a hands-on approach. After the Immersive Experience phase, we go back into the Critical Reflection phase again so determine if there are other areas where we are not meeting our customer's expectations.

How does your organization see failure? Does your organization see failure as the opportunity to learn or the roadblock that gets in the way? The Fail Fast model can help Agile be more productive in your organization by changing the way your organization views failure.

Failing and learning plays a prominent role in many of the principles including "Continuous attention to technical excellence and good design enhances agility." We think it's pretty safe to say that you can learn technical excellence by having immersive experiences that are hands-on with the product. These immersive, hands-on experiences with the product allow you to learn more about the product. The experience also brings about several failures when you try to perform a function with the product, and it doesn't work that way you expect. You learn from these failures and why they happened allowing you to avoid them in the future. However, they also serve as ways to build better user experiences. You can take what you learned from the failure and bring it forward into new and exciting designs.

There are more examples for each of the principles. Spend a few moments to look at the Agile Principles and see how those principles can be realized using failure and learning from the failure. Come up with a list that you can take back to your team and discuss how to apply them.

Successful Project Factors

In a 2018 study called "The Chaos Report" by the Standish Group found the top 10 factors of successful projects.

Top 10 Project Success Factors
1. User Involvement - 15.9%
2. Executive Management Support - 13.9%
3. Clear Statement of Requirements - 13.0%
4. Proper Planning – 9.6%
5. Realistic Expectations – 8.2%
6. Smaller Project Milestones – 7.7%
7. Competent Staff – 7.2%
8. Ownership – 5.3%
9. Clear Vision & Objectives – 2.9%
10. Hard-Working, Focused Staff – 2.4%

All Others – 13.9%

User involvement hits the top of the list and strongly reflects the desire to have technical and non-technical individuals working together to form better solutions. This allows the team to fail and learn from that failure more quickly.

Executive sponsorship hits the number 2 spot of the list. Executive Sponsorship is referring to support from senior managers and top executives. Without this executive support, it is difficult to get a project completed successfully. At times always getting executive support can make your organization feel top heavy and slow to respond to changes in your industry. It can also mean senior management contradicts your team decisions at times. To effectively make Agile and the Fail Fast model work effectively in any organization, you will need to have strong executive support as a part of the equation in getting your organization to adopt Agile and the Fail Fast Model.

Now here is one that should not be a surprise given the nature of this chapter. "Smaller Project Milestones" is another way of saying Agile or smaller iterations. Which means Fail Fast and Fail Safe.

All of these things on the list above have to work together to change the organization's culture. We believe that to improve project success, your organizational culture needs to embrace failure as a way to learn, utilize the Fail Fast model, and break old habits of not creating environments that allow failure to occur quickly and safely. Agile practices help.

Learning and failure also play a significant role in these organizations. Learning quickly supports individuals in being more fluid. Quickly learning requires the immersive experience phase as outlined in the Fail Fast model. These immersive experiences lead to critical reflection on how to improve. Conceptualize Improvements takes the immersive experiences and critical reflections to created proposed solutions for improvement. Again, Agile practices help.

If you are not convinced that Agile and the Fail Fast model should be a part of your organization, let's take a look at statistics from the Standish Group study on the percentages of features that are or are not used after using a Waterfall development approach.

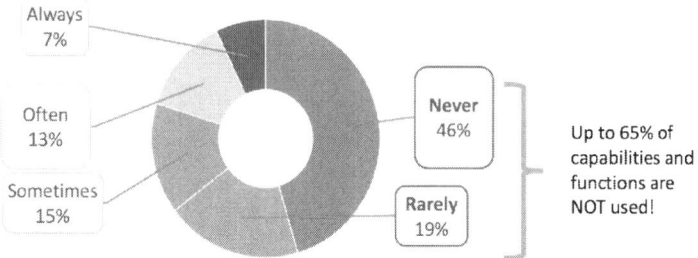

Think of the waste! 46% of capabilities and functions are NEVER used. By the time the actual release date comes, requirements have changed, the business value changed, the market changed, too many things have changed. As a result, almost half of the development work is useless. Much of this waste can be eliminated by failing fast and learning from that failure. Capabilities and functions are no longer used in most cases because they were not designed with the business process in mind. By having greater user involvement, we ensure failures in meeting user or customer expectations are quickly brought forward where we can take the failure, critically reflect on it, and engage in conceptualizing improvements to redesign the capability and function so that it does meet user and customer expectations.

We understand the principles of Agile and Fail Fast will be hard to implement. We understand that your company may not be able to do all of them. As an individual, can you be more agile?

Agile as a Mindset

Now we get to our next definition of "Agile is a mindset." The Agile mindset is defined by your values, guided by the principles of Agile and delivered through an Agile practice (methodology). This goes back to our discussion on right brain/left brain. If you think about the principles, they are right-brained. They guide us to the do the right thing. How do we get it done? Left-brained approaches that include Scrum, Kanban, and other Agile Methodologies.

The future is here, and Agile and DevOps are forcing us to rethink how we work.

A brief note about DevOps. DevOps comes from "development" and "operations" and is a software engineering culture. DevOps proposes the unification of software development and operations. It advocates automating and monitoring processes that have historically been manual or slow. DevOps means that teams are no longer "siloed," and the development and operations teams are merged into a single team where they work together across the entire application lifecycle. This process creates and demands "fluid" roles. Fluid roles are not limited to a single function. Teams will bring in specialists as needed. More flexibility and adaptability are required in DevOps environments.

Everyone wants to get it done faster, less is more, and do more with less. The mistruth is that Agile and DevOps are a cure for all our problems and this is wasting a lot of time, effort and money.

We (and many colleagues) have seen clients spend hundreds of thousands of dollars converting office space into Agile and DevOps friendly working environments based on the idea that it was the right approach. A lot of new office furniture was sold in the belief that by merely transforming workspaces, the individuals in those workspaces would be transformed is crazy.

The new space looks great, but furniture can't fix a failed process. The ancient sage advice holds true, "A fool with a tool is still a fool if they have not been taught how to use it." The outcome was simple. Individuals felt they got a beautiful workspace, but the process was broken, work was frustrating, and nothing got done.

A better approach is to experiment with a smaller group first. Let's find out how this will work before rolling it out company-wide. Micro iterations are a part of the Fail Safe model and allow failure to occur more quickly so you can conceptualize improvements. Let the demo run for 2 – 3 months. Let the experiment fail. For example, the failure was that the tables didn't fit into the cubicles well and individuals kept backing their chairs into those tables. Critically think through the failure. What is happening with the tables and chairs? Try to get to the root cause. When the cubicles were designed, this wasn't foreseen as a potential failure. You can't make beneficial changes to solve the problem unless it is fully understood. Move on to conceptualize improvements. What are your options to fix the problem? Do we need to change the furniture layout?

When you perform viability analysis, determine if you are ready to start converting more cubicles to the new format.

Nothing is a greater mistruth then the belief that a popular or new tool or methodology will fix all the problems you face.

You and your organization recognize that because technology is changing faster than our methods of delivering new technology that it's getting more and more challenging to keep up, so we try looking for quick fixes to solve our challenges. Carefully stepping back and experimenting with DevOps and the 18 different types of Agile will produce better results. By experimenting and failing, you learn from the failure and can make improvements. We have to stop listening to the voices that say we must do Agile purely and without modifications. This type of thinking stops you and your organization from experimenting and learning the best approach that works for your organization. You might even develop the 19th version of Agile.

The way you deliver business value must be critically reflected on, new improvements need to be conceptualized, and changes must be implemented on a routine basis. That is not able to happen if you don't embrace failure and learning from the failure. Failure is the opportunity for improvement.

Agile is not as Agile as we think it is. Does Agile work with Artificial Intelligence projects? Data Science? We don't know if Agile or DevOps will work with them unless we are willing to experiment, fail in that experiment, learn from that failure and transform ourselves to avoid that failure in the future.

Let's take a look at one specific example. Looking at the table below we can see the new technologies that have emerged to put pressure on the organization for change. In this example, it shows how advancing technology has impacted the Business Analyst role. Once upon a time, Agile purists said, "Get rid of all the Business Analysts!" Then they said, "Just get rid of the title." Then they said, "Well you can have a Business Analyst on a team." Now we see that the impact of technology is not making Business Analysts go away, it is creating many new roles where just a few existed before.

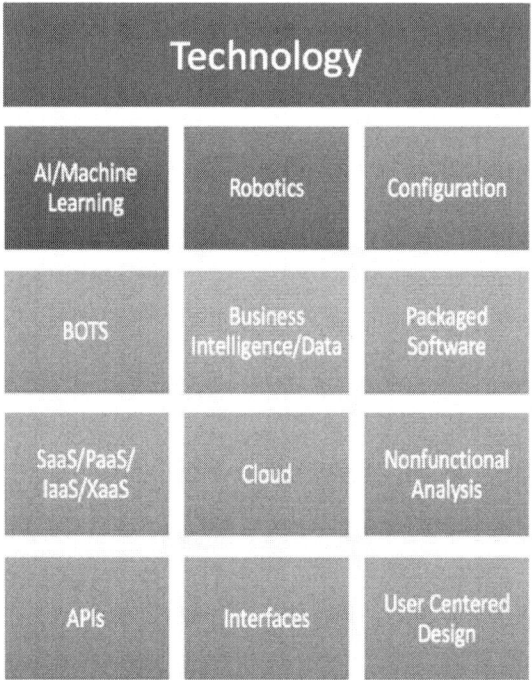

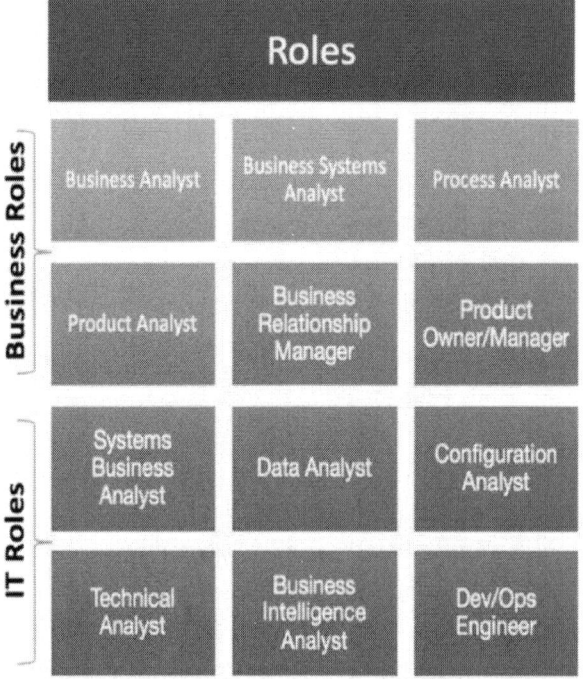

How will you experiment with these technologies? How can you understand how they will impact the roles within your organization?

Design experiments that can test the new technology and the new roles to support it. Let's say you are moving into the Artificial Intelligence space. What experiment could you create to determine if AI will work in your organization? Once that experiment is devised, the next step is to determine how the individuals and their roles within your organization would be impacted if this experiment is put in a production or customer-facing environment. Do you have the processes and roles established to "operationalize" AI?

Perform the experiments with micro iterations. Focus on something small enough that you can learn from it but not too big that it would be challenging to conceptualize improvements. This immersive experience of experimenting will allow your team to learn while engaging with the AI technology.

Once that immersive experience with the AI technology is finished then critically think back on the immersive experience. What failed? Get to the root cause of that failure and then move into conceptualizing improvements to the experiment so failure will not occur.

Lather. Rinse. Repeat. Keep experimenting and learning from failure to make improvements. When the experiments are successful, add a slightly larger experiment on top of the experiment you just finished. Follow the same process. Keep it up adding small components on top of each other and making sure you are building on a solid foundation. As the experiments progress, you will at some point be able to release your experiment into a customer-facing or production environment.

Learning from failure is both about the experiment failure and the roles that supported or performed the experiment. Roles will fail just as easily as a process. Adjust roles as needed to make the experiment runs more effectively and successfully. Training and hands-on experience are a powerful combination to build up the skill sets of roles and individual in your organization. Many organizations are taking the stand that there are no roles and they are simply calling their people "Team Members." They do everything – or at least that is their hope. This too is a trend and like all of the additional roles we see being created, they come from experimentation. In the end, do what is best for your organization. Just remember not to throw out the business analysis with the bathwater.

Not all technologies are designed to be Agile or DevOps. The myth that you need to select only one methodology is false. A methodology is a tool. No two projects are the same. If every project is different, then why are you using the same tool for each of them? Bob and Paul both have been on projects where multiple methodologies have been in play on multi-million dollar, cross-functional, company-wide projects. Scrum, Kanban, XP, Waterfall Incremental and homegrown; same project, 5 methodologies.

Think about this in terms of a woodworker. Woodworkers are building you a new kitchen table. If the only tool they can use is a hammer, it's going to be pretty tough to make that table. It most certainly isn't going to be pretty. Have multiple methodologies available to your teams for use allows them to think about what they as a team need to operate best. Also, it will enable your teams the ability to self-direct and figure out if combing methodologies will help them perform more effectively. One of the principles of Agile is to let the team learn from their behavior and adjust.

How would a team choose a methodology or methodologies for their project? There are a few things to consider before choosing a methodology:

1. Look at the capabilities your project will deliver. These capabilities most likely require infrastructure like hardware, software, networks, and more. Capabilities need an

architecture that will support them. In the Chaos Report, standard architecture was one of the factors for successful projects. How complicated is the capability? What are the impacts to the organization with this new or enhanced capability? Do we currently have the infrastructure to develop this capability? What environments will you need to run your experiment effectively?

2. What techniques are required to elicit and analyze the capability? How will you design your experiments?

You may not be able to answer these questions adequately initially. Research and answer them as best as you can as this information is used to help you determine the appropriate methodology or methodologies.

Pick your methodology intentionally instead of just picking Agile or Waterfall. First, determine if other methodologies will help you to develop and perform your experiments or development activities more efficiently.

Another item to consider in picking a methodology is some new approaches that call themselves Agile are Waterfall overlays for Agile. We don't have any grudges against these methodologies but ask you to keep in mind your vision for what you want from a methodology before selecting a methodology. Don't pick the methodology based on the latest trends. Many trends are commerce. Think about it.

Projects like a mobile application, automation of customer forms, website development, e-commerce, and client facing screens or reports are going to work great in Agile.

A waterfall-incremental methodology would be a better choice for projects that are heavily focused on integrating a large number of systems, data migration, has significant regulatory or legal implications, and large construction type projects.

If your project interfaces with a large number of other Agile or DevOps projects, you may need to utilize Waterfall because not all Agile and DevOps teams will be able to deliver on a coordinated schedule. Some teams have different heartbeats or sprint durations. One team can have a sprint for two weeks, and another has a sprint of one month. Other teams might be utilizing Kanban and not have a distinct timeframe for delivery. A good example is an organization that has multiple Kanban teams. One team works on 4-5 things at a time and another only works on one thing at a time regardless of size or duration. It takes a village.

A brief word about your future with Agile and DevOps. Many companies look up to Amazon, Microsoft, and Google when they think about Agile and DevOps. Keep in mind they have invested hundreds of millions of dollars in their systems that greatly assist with their ability to structure themselves as strong Agile and DevOps teams. The investment is in the development of your systems, continuous training of your people (all the time and not just once a year) and tools that people need to get their jobs done. That kind of investment will make your Agile/DevOps journey easier.

Be a More Agile You - Agility

What about just you? What if your company is not ready to commit to Agile and DevOps just yet? Can you be more Agile? Yes, you can! Do you have a plan? Yes, you will! You can indeed live by many of the principles, and in turn, you will become the role model for your organization. Here are a few tips for being personally Agile:

- Rather than send that long email, call them or stop by their cube and talk to them. Email is only good for one thing, and that is a CYA move (please don't make us spell that acronym out). Emails should be about quick communication – five sentences or less. If emails are getting longer than five sentences, talk in person. That aligns with the idea that colocation is better (principle #6). We value direct communication getting answers quickly and efficiently.

- When someone is unhappy with the timeframe, ask the question: "What can I get you right now that will add value?" This aligns with principle #1.

- Trust aligns with principle #5. Start trusting people to do their job and they will trust you. Help build a culture of trust. If you trust others and they are not trusting you ask why? If you do not trust others and they trust you ask why? Then sit down together and talk about how you can get there. Trust is vital to make any of these principles work.

- Take time for self-reflection as it aligns with principle #12. We are all told that self-reflection is helpful, but many of us do not follow this critical principle. We need to evaluate our experiments, our tests, our failures, and our successes to determine course corrections. Reflect on what you have done to demonstrate how reflection can be used as a team to improve continuously. Critical reflection is also part of the fail fast model.

- Adopt Kanban as a way of life. See "The Rule of 3."

The Rule of 3

Your rule of 3 is about prioritizing what you do every day to ensure you are working on the most critical tasks that offer the most value. There are many other rules of 3, but for this example let's focus on your To Do list. Here is how it works:

- Start by creating four columns on a piece of paper entitled:
    - Backlog
    - To Do
    - Doing
    - Done

    You can do this with the spreadsheet application of your choice if you wish. Bob uses the notes function on my phone. You may decide at some point that you want to change the titles of the four columns at some point. Experiment and see where it goes.

    What happened to the rule of 3? Hold on; we are getting there.

- Next, create your list of wants, ideas, wishes, known to-dos and whatever else you think this list should include and put them in the backlog column/list. It should be an extensive list. In Agile, the backlog contains all of the requirements your business partners want to be developed. In your world, it is all of the things you want to get done whether it is picking up the kids from band practice, remodeling your kitchen, or a present you need to buy for a birthday. You can be as detailed as you want to be.

- Now prioritize all of those 'backlog' items and only choose three that are your highest priority. Maybe it is due to timing, or maybe it is due to the value it brings you. Move them from your backlog to your to-do list. The 'To Do' is not doing them. It is preparing to do them. Analysis, preparation, research, study, anything to get ready.

- When you have prepared appropriately and are ready to do one task, move it to the 'Doing' column.

- Once you move something from your 'To Do' to the 'Doing,' you may now move something from the 'Backlog to your 'To Do.'

- When you have finished a task move it to the 'Done' column to show your success. Of course, now move something from 'To Do' to 'Doing.'

The key to the rule of 3 is that you cannot have more than three items in your 'To Do' or 'Doing' columns. You can have as many as you want in the 'Backlog' or your 'Done' columns.

This process helps us focus on what is important and what brings value to our lives. Our brains often wander – "Squirrel! Shiny! Glittery!" or we become scattered or frustrated. The rule of 3 can take your wandering squirrel and keep you on track delivering value. Now think about the role model you will become for others!

Keep in mind you make this the rule of 5, the rule of 9, or any number that works for you.

What other ways can you be a more Agile you? Take some time to create a list of things that might work for you. Then put them in your backlog to work on!

Wrapping Up Agile and Failure

You will be Agile as you can be for where you are today. Recognize you will evolve. There is a reason why people say we are "Agile-ish," "Scrum but…," "WAgile," "Scrummerfall" or other variations. You don't need to say you are Agile. You need to embrace that you are finding better ways of delivering and that means you are experimenting, failing fast and learning from those failures.

Use Agile and DevOps as a means to an end. A year from now you will look back and say, "We thought we were Agile but not really. Our experiments have helped us improve greatly." A year from then you will look back see your efforts have allowed you to be even more flexible and adaptable.

Recognize it is not about IT transformation. Stop saying "Let's let IT put their ducks in a row about transformation before we start experimenting or transforming." Transformation can't happen until the entire organization is working together to transform.

Yes, Agile is about an iterative approach to software development, but it only works when your business partners play along. Guess what? The Business needs to be Agile too. Agile works great on non-IT projects.

Do what is best for your organization. Do what makes sense. It's okay if parts of your organization are going to be Waterfall and other parts of Agile. One part of your organization may have created their own methodology and called it Pickles (which opens up a new topic that we will not get into, called Gherkin – a business readable, domain-specific language that lets you describe the software's behavior without detailing how the behavior is implemented). The important part is to choose a methodology that will allow faster failure, critical reflection on the failure, and learn from that failure. Go ahead name your methodology "Twinkle." Think about an entire organization saying that word over and over. Makes you smile, doesn't it?

Chapter 11

Putting Fun in Failure

"People rarely succeed unless they have fun in what they are doing."

- Dale Carnegie

Failure is fun! Really? Well, it can be, and it should be. Failing is stressful enough, even when you are used to it, so why not put some fun in the process? There are only so many times you can sing "Let It Go" to relieve the stress. Singing that song won't help all of those parents out there that have heard that song 500 times as their children keep watching Frozen again, and again, and again (and trust me they don't want to hear it again). Injecting some fun into our work and especially into our failure will help. While the idea of having fun while we work is not necessarily new, embracing it in corporate environments to drive success and failure is a different story.

A focused approach to incorporating fun is fairly new and is based on a concept called gamification. Gamification? Isn't that just playing video games? Wait, is that old school board games? I know, it's Candy Crush on my phone until 2 AM! We have been asked the question again and again "Why would we do that at work? We just need to get stuff done. Gamification is not for my team. That sounds like a waste of time." Gamification is perfect for teams of all sizes, backgrounds, and industries. To make gamification work, we must structure it with some thought. The term gamification often strikes fear into people and the irony is that it is intended to do the exact opposite.

The dictionary defines Gamification as:

"The application of typical elements of game playing (e.g., point scoring, competition with others, rules of play) to other areas of activity, typically as an online marketing technique to encourage engagement with a product or service."

Engineering circles define gamification as "Gamification is the implementation of game-based thinking or game mechanics to engage users and solve problems." [65]

Another gamification definition is "Instead of creating full games, gamification's guiding idea is to use elements of game design in non-game contexts, products, and services to motivate desired behaviors." [66]

The idea behind gamification is that it will introduce fun into the work environment making it seem that work is not a **"must have"** but rather an **"I want to."** Turn the fear into fun!

The Greek historian Herodotus[67] was known as "The Father of History," and he produced just one major work entitled "The Histories." In this work, he documents the first known example of a game, which may surprise you in that it was not for people's entertainment. The game was created in the kingdom of Lydia, who was in peril from a great famine. The strategy was to divert people's minds from their hunger, and the other problems of the day. The game was believed to be the first form of dice made from sheep's knuckles. This was the first known game with "props" or "tools." Now, Herodotus stated he repeated a story that he was told, and he could not verify the story. Other examples of gaming might precede

---

[65] WikiPedia, en.wikipedia.org/w/index.php?title=Gamification&oldid=854161458, Gamification, Zichermann & Cunningham, Various Contributors, August 2018
[66] Adrianomescia, adrianomescia.com/deterding-defining-gamification/, Deterding – defining gamification, Adriano Mescia, July 2015
[67] Wikipedia, en.wikipedia.org/w/index.php?title=Herodotus&oldid=858926727, Herodotus, Various Contributors, September 2018

the Lydians in China and Egypt. Gaming has been around a long time and has had multiple uses.

Bob grew up in a game playing family. We wanted to have fun as a family, so we made games out of our chores and simple activities or making games with random things we could find a purpose for in our amusement. When you grow up poor, it is what you do. We played games when there was nothing else to do. Atari? Not in our household! Bob wanted his MTV, but that didn't happen until more than a decade later. There was a lot of catching up on music videos.

Consider that game theory has been used in our society for generations. Can you think of a group or two that has been using it for 100 + years? If you guessed Scouting programs (Girls and Boys), you would be right. Scouts are challenged with mastering various disciplines. These challenges combine the ideas of fun and hard work to gain mastery of several disciplines. When Scouts are completely immersed in learning, they are completely engaged in the game. When Scouts are completely immersed in the learning, they are motivated to continue to work this way. They are motivated because they can see progress – they are rewarded (remember the badges?).

Being fully immersed and engaged in training reinforces our company motto of Think, Learn and Work Differently.

Marc Prensky[68], author of Digital Games Based Learning states:

- Games are a form of fun. That gives us enjoyment and pleasure.
- Games are a form of play. That gives us intense and passionate involvement.
- Games have rules. That gives us structure.
- Games have goals. That gives us motivation.
- Games have outcomes and feedback. That gives us learning.
- Games are adaptive. That gives flow.
- Games have problem-solving. That sparks our creativity.
- Games have interactions. That gives us social groups.
- Games have representation and story. That gives us emotion.

These are all key components of highly functioning teams. If teams can embrace game play along with failing fast and safe, you will have some pretty amazing people working for you or you will in fact, be that amazing person on the team!

Your Brain on Fun

Still not convinced? The founder of Dopamine Inc, Gabe Zichermann[69], references a study that finding a way to make work fun increases an employee's ability to retain skills by 40%. Also, microlearning helps the learner remain productive. Now combine the idea of gamification, Agile and failing fast. Not only do you create a

---

[68] Wikipedia, en.wikipedia.org/w/index.php?title=Marc_Prensky&oldid=836741173, Marc Prensky, Various Contributors, April 2018

[69] Wikipedia, en.wikipedia.org/w/index.php?title=Gabe_Zichermann&oldid=846775897, Gabe Zichermann, Vairous Contributors, June 2018

robust learning environment, but you also create a higher skills retention rate.

Certified neurologist Dr. Judy Willis[70] has done extensive research proving out that dopamine is released during tasks where people are unsure of the outcome, or they are enjoying the task. Guess what games do to us? Games are unpredictable, and dopamine is released in our brains.

We incorporate gamification into our style of teaching as much as we can. We experiment almost every class with a tweak to something or create something new.

We often get questions from students during class exercises ranging from "Are we going to get the answer for this exercise?" or "How do we know this will work?" We don't directly answer those questions because we know that our goal is to utilize gamification into their training. We create a sense of uncertainty in how to accomplish the task which creates tension. There is a sense of urgency with students because they think they are expected to succeed in class (even though our expectation is that they fail fast). Students seek out advice quickly trying to remove the tension, but the advice only increases the tension. You know, those "it depends" types of answers. Then, add the pressure of a time constraint what do you get? Dopamine! There should be a song for that. "Dopamine, dopamine, dopa, dopa, mine… la, la, la." You have to challenge your skills. If your skills are not challenged, you will become bored and stop caring. We let our students fail fast and fail safe so that they remember the exercise, the

---

[70] Psychology Today, www.psychologytoday.com/us/experts/judy-willis-md-med, Judy Willis M.D., M.Ed., August 2018

advice, and the moment when everything falls together. Then the lightbulbs go on.

It is time to stop looking down on having fun at work and failing. It is time to embrace it and see what new success the fun and failing will bring you.

"Bob and Paul, I think I am starting to understand that combining fun and failing are good things. What are some ways to introduce a little fun into the work environment?"

Glad you asked. What do the following four items have in common?

The Marshmallow Game

Let's first start with a well-known example; the Marshmallow Challenge. Peter Skillman of Palm Inc. invented the Marshmallow Challenge, and Tom Wujec of Autodesk popularized it. Often people use this challenge to learn about team bonding, leadership, prototyping or planning. We have our twist on it which is, of course, helping teams fail fast and safe. Failing fast and safe, and learning from that failure is key to building the best structure. Here is how it works.

1. Form teams of 3-5 people per team.
2. Each team gets the following:
   a. 20 sticks of spaghetti.
   b. A yard of string/twine. Bring scissors – it's a game changer when you forget it.
   c. A yard of tape (I usually use painter's tape to make it more difficult).
   d. 1 standard marshmallow. Bring extras – people will eat them.
3. Each team has 18 minutes to build the tallest tower with the supplies provided. A marshmallow must be at the top of the structure.
4. The structure is free standing – no hanging from the ceiling. You can cut the string, break the spaghetti, rip the tape, tape things to the table/surface, tie the string, tape the string, but you must leave the marshmallow "mostly" intact. It can be pierced but in no other way can it be separated or broken apart.
5. When the timer goes off (18 minutes), all hands must be off of their structures. If they fall, they fall. If they stand for more than 10 seconds after the timer, they qualify to be measured. Tallest structure wins. You choose the prize. And yes, if the spaghetti pierces the marshmallow and can be seen higher than the actual marshmallow the team is disqualified.
6. Now have the discussion about why failing fast and safe is important. Discuss the need to micro-learn from their tests.

Try it, test it, fail it. Learn. Try it again, test it again, fail it. Learn. Did anyone test the marshmallow on a piece of spaghetti right away? Lots of teams do not test that approach. With a couple of minutes left on the timer, they are scrambling when they finally realize they failed too late. Who does the best at this exercise? Architects are the best at this exercise. They understand structure. Who is second best at this exercise? Children just out of Kindergarten! They experiment. They are comfortable with failing fast and failing safe! As adults, we bring our baggage to the table with us that gets in the way of failing fast and safe.

7. Six months later, get the teams together again to do the exercise. However, put a twist into it. Use an oversized marshmallow that is much heavier. Give them twice the amount of tape but half the spaghetti. My favorite is to add a completely new element with unusual rules that they don't see coming. Wild cards add unexpected actions to the game. In game play, this always provides the fear of the unknown (dopamine here we come!). Wild cards can be anything you want them to be.

    a. Steal two pieces of spaghetti from a team.
    b. Get an extra minute.
    c. Penalize a team with a lost minute (or two, three…).
    d. One team must use two marshmallows at the top of their structure.
    e. Just get creative!

8. How did the team do this time? Have they learned about failing fast and safe? Are they practicing it? Is it showing in their work?

"That sounds like fun guys! What else ya got?"

Brainstorming

Everyone knows how to brainstorm right? Well, they probably do, but they are most likely using an older technique instead of more modern brainstorming technique.

Our friend Wikipedia says: "Brainstorming[71] is a technique for group creativity by which efforts are made to find a conclusion for a specific problem by gathering a list of ideas spontaneously contributed by its members." In other words, brainstorming is a situation where a group of people meets to generate new ideas and solutions around a specific domain of interest by removing inhibitions. People are able to think more freely, and they suggest many spontaneous new ideas as possible. All the ideas are written down and are not criticized. After the brainstorming session, the ideas are evaluated. The term was popularized by Alex Faickney Osborn[72] in the 1953 book "Applied Imagination."

Our goal here is not to teach you basic brainstorming (don't trash people's ideas, timebox the effort, etc.) but rather to show you how it can be more fun! Let's mix it up a little.

Here are a few fun examples of where brainstorming can be kicked up a notch.

Brainstorming: Super Heroes

Super Hero brainstorming is where each participant picks their favorite superhero. Whatever comic book universe they want to choose is fine. Let the Marvel and DC universes clash in all their glory. It gets crazier when someone adds in Star Wars and Harry

---

[71] Wikipedia, en.wikipedia.org/w/index.php?title=Brainstorming&oldid=854993489, Brainstorming, Various Contributors, August 2018
[72] Wikipedia, en.wikipedia.org/w/index.php?title=Alex_Faickney_Osborn&oldid=809869611, Alex Faickney Osborn, Various Contributors, November 2017

Potter). Here are some examples:

- Mary chooses Wonder Woman
- Jon decides to pick Thor
- Kat would like Shuri for her hero (okay this might be newer for you – she was the so-very-awesome Tech Guru in Black Panther)
- Zach selects Spiderman
- TK chooses The Hulk

Now all participants must brainstorm from the perspective of their superhero and their superpowers. How would we solve the problem with Wonder Woman's lasso of truth? With her invisible plane? How could we make it stick like Spiderman's webbing? How could we smash it apart like The Hulk? This style of brainstorming works incredibly well when folks embrace the idea that having fun can work to yield results. This is one of my favorites. Better yet, get folks to do the voices or imitation of the hero when they present their ideas. "Hulk SMASH!"

Brainstorming: Iconic Figures

A twist on Super Heroes. Although Super Heroes is one of my favorites, not everyone can always relate. Instead, try iconic figures. Rather than do Wonder Woman or Thor, try Madame Curie or Albert Einstein. The possibilities are endless. We allow people access to their computers with this version so that they can look up the qualities, characteristics, and skills of the iconic figure. We have also done this where we have chosen the iconic figures in advance and put them in a bowl that people draw randomly. They must also act out the character if they can! There have been some pretty impressive Albert Einstein and Marilyn Monroe impressions.

Brainstorming: Time Travel

Rather than solve the problem at the moment, why not get perspective? Let's get into the DeLorean or your TARDIS and time travel to a different year! Now if we have to explain Back to the Future or Doctor Who…

Here is how it works.

- Choose a period in the past, say, 5 years ago. Solve the problem like you would have solved it 5 years ago. To do this, you may need an expert from that time period.
- Now solve it as you would today with your current resources.
- Finally, choose a time period in the future – maybe 10 years from now. Imagine what the technology will be like, what the future may hold.

What this style of brainstorming will do is provide perspective. It gives you knowledge from the past. Keep any past knowledge that is valid. Jettison some of that past knowledge if it is faulty. Ask the questions, "What if…" to create the future. This is an excellent approach when comparing current and future states of systems. Bob usually makes people draw their time machine. Kudos to the person that drew the H.G. Wells time machine in full detail by the way – you know who you are, and it was amazing! Drawing your time machine is a great way to have some additional fun.

Brainstorming: Pictures Only Please

This one is simple. They brainstorm in pictures only. Think of it as drawing Charades. Draw the idea in a picture to make it more visual. Then everyone guesses what the idea is. It takes a little more time, but for one workshop it really helps teams to relax and think about how they communicate. Turn it into a competition. Points for who guesses the drawing first. Then you can also do it in the style of the game "Concentration." Have them draw first (without showing it to anyone), then put stickies over the drawing and remove them one at a time to see who gets it first.

What creative ways can you think up that will liven up the workshop brainstorming? Currently, we are up to about 75 different versions of brainstorming. Now we are sure you are saying, "Why do you need so many?" The more techniques you know, the better chance you have filling a specific need. Every situation is different. Every team is different. Our approach is to keep finding ways to help people have fun and failing fast and safe.

Think about what you can bring to the table when you are designing your tests, your opportunities for failure. Can you gamify them? Pull from your existing knowledge of games. Which ones might work? We have used aspects of all of the following in our work:

- There are board games: Monopoly, Life, Chutes and Ladders, Concentration, Wheel of Fortune, Pictionary, Uno, Hopscotch, Darts, Chess, Dominoes, Jenga, Trivial Pursuit, Clue, Risk, Tic-Tac-Toe, Connect 4, Scattergories, Balderdash, Flipping A Coin, and Rock/Paper/Scissors or its variant Rock/Paper/Scissors/Lizard/Spock.
- There are all sorts of card games: Poker, 500, Gin Rummy, Hearts, and Spades to just name a few.

Are you ready to put the fun in failure? We are!

Work should just be a lot more fun!

Chapter 12

Think, Learn, Work Differently

It's time to tell the tales of failures of the past. We pulled out the more humorous of the stories we could tell of our past failures to see how we should have applied the fail fast and fail safe models.

There are so many stories of personal failure that could have been told. Bob's near death experience getting hit by a motorcycle and being paralyzed for 36 hours. Bob's near death experience with cracking his skull open on a bolt that holds down the seats in a theater. Bob's near death experience falling off of 20 foot tall slide. Bob's near death experience in isolation for 36 hours and being treated for Ebola. How is Bob even alive anyway? Bob does think his near death stories are all quite funny now. Paul smashing into a pine tree during his first ski lesson. Paul getting bit by a 250 lb. Mastiff less than two weeks before getting married. We'll save the rest of the stories for the sequel. Whether it has been work or personal failure, it has led us to believe one very powerful truth; we must Think, Learn & Work Differently.

Freight Elevators

We were engaged on an infrastructure upgrade project very early in our careers. This project entailed installing 50 server racks in a large server farm. The server farm was located under the building in sub level 3 for security. We ensured that the specifications for the servers would meet the processing demand we were expecting, validated the server software would work, and architected each server for the best performance. We then set out focusing on the electrical side of things by ensuring the servers would have enough circuits and outlets. We worked with the environmental control team to make sure the cooling and air flow in the room would support the servers without overheating. Stress tests were conducted on the network to validate the heavy traffic loads wouldn't collapse the network. The backups and restores for disaster recovery were thoroughly tested.

We worked with the vendor on designing the racks and came upon the idea to have the racks assembled, and the servers bolted into those racks prior to delivery. This would allow the vendor to perform most of the work outside for the server room and it avoided performing any construction or assembly around production servers. We made sure the pathway from the receiving dock all the way to the server room was clear of boxes and equipment.

The first rack arrived on the receiving dock and was in perfect shape. The truck took a slow journey along very smooth roads with little traffic. We unloaded the server racks and wheeled it to the freight elevator. The freight elevator doors open, and we started to wheel the server racks into the elevator.

The server rack didn't fit.

After every possible way an object could be put into a freight elevator was attempted, we couldn't get it to fit. The rack was too tall. We tried removing the ceiling of the elevator, but that was bolted down. We tried removing the rack from underneath the server rack but it was still too tall. We removed all the packing materials and it was still too tall.

We were finally forced to remove all the servers from the rack and disassemble the rack entirely. Then carefully put everything in the freight elevator piece by piece and bring it down to the server room where it had to be assembled all over again. How did the other servers in the server room get in there? They brought everything down piece by piece. We missed that fact. It was a costly failure and the project finished late.

Who would think a freight elevator wouldn't be big enough? We would have figured that out had we used a cardboard box that was life-sized and attempted to use the freight elevator. Our assumption that it would fit was a bad one. We tried to take a lesson learned from the last team that put in new servers, but we failed because we missed some of the important details. We didn't experiment to confirm our assumptions. All the best planning can be laid to waste by an unexpected event. We certainly didn't fail fast or safe.

Experiment to validate your assumptions. Never leave assumptions laying around unvalidated. Always fail safely by validating the environment – or in this case the size of the freight elevator. When we performed server additions in the future, we always experimented to get an immersive experience even if it meant looking strange moving a cardboard replica of a server rack around the hallways.

Experimenting and test runs not only finds potential failure points but also allows us to critically reflect on making the process better and more efficient prior to the actual implementation. Fail fast and learn fast.

Endless Formulas

We had a client that was having problems with one of their agile projects. The project was to create a quality score by putting manufacturing data through a formula that would produce a score used to determine the quality of the product. The project was running for two years and had yet to finish.

We quickly discovered that the formula was continually being changed due to standards being modified by the international standards body.

After spending six months to develop, validate and test the formula to ensure it will meet the standardized measurements, the technology team discovered the calculation or results were incorrect. The science team never interacted with the data to refine the formula.

The solution here was that the science team needed a way to interact with the data more directly. They needed to play with the data to understand it. The science team didn't critically reflect on the data at all until they got the finished report from the technology team. Since the technology team took 3-6 months before delivery of the finished report, the international standards body had already updated the standard before the new report was delivered.

We stopped the project entirely and went about redesigning a new solution. We broke the formula up into key components. We then decided to create an interface that would allow the science team to change those components in response to the international standards changes. This also have the team the ability to play with the measurements and produce results on a realistic set of data. We abandoned the idea of hard-coded formulas. We created multiple environments for each member of the science team that were easily reset. Reset means returning the environment to an exact copy of the production environment. This allowed the team to be even more flexible in playing in the environments because if they changed the measurements too extensively, they could reset it back to the same standards in production. Since the environments were safely disconnected from production, we didn't need to fear making production errors.

Then we put in place processes and methods for moving configuration from their individual test environments into a common test environment. We delivered a configurable system within 60 days.

By creating spaces for experimentation and failure to occur, we were able to help the teams critically reflect on the data. They could adjust the data using 42 different metrics that were captured through the manufacturing process and then apply them using realistic data. Additionally, the final formula itself was configurable so the result based on the 42 metrics could be changed. The science team was able to reflect better critically and was then able to conceptualize improvements.

Think Differently

Our company motto is Think, Learn, & Work Differently. We believe that you must think differently to learn differently. This an essential step in failing fast and failing safe. By failing fast and safe, you will then learn differently. You will learn in ways you have not learned before! Once you have learned differently you will go in directions you did not know were possible. You will finally work differently. It is a powerful concept, and we believe that you will succeed if you are ready.

Being Flexible, Adaptable and Open to Change

Changes are the opportunity to learn something new and to take on a new challenge where you can grow. Change allows you to experiment and fail to learn.

- Over your lifetime you will be exposed to many opportunities. Google, Amazon, and Apple employ thousands of people who have careers in things that didn't exist 18 months ago.

- In the next ten years what you know now will have changed to something new because technology evolves in 3 to 5-year cycles.

- You are a unique and multifaceted person that can't be matched to just one role and one industry. Where you are career-wise today, will be very different in the future.

Potbelly started as an antique store in 1977. At one point, the owners decided to serve sandwiches to help draw more traffic into the store. The customers proved to be more interested in the sandwiches than the vintage glass doorknobs and Tiffany lamps. Potbelly started selling only sandwiches and became a chain of restaurants - that's a pretty significant change!

Change can take you in a completely different direction. Being adaptable and not allowing yourself to be locked into a certain mindset helps you see innovations. As you experiment with concepts and ideas, you will need to be adaptable. The outcomes of experiments may lead you in a whole new direction you didn't expect. Product design and capabilities will change very fast when you fail fast. We need to embrace the mindset of flexibility and adaptability to fail fast and to fail safe.

When you are inflexible and have a constrained mindset, a failure can be challenging to learn from. If you were an antique store that discovered you were better at sandwiches, would you take that leap to become a sandwich shop?

Thinking Differently Self-Evaluation

So, you have been through most of the book now. Are you thinking differently? Let's see. Score the following statements below based on how you look at yourself. Add the value to the score box and then total up all your answers (there are two pages). Do not be influenced by looking at Step 3 Understand Your Score first (spoilers!). Do this in the order provided.

Scoring System

Score yourself on a scale from 1 to 5. 1, 2, 3, 4, or 5.

Score	Definition
1	Never, rarely, or less than 10% of the time

2   Infrequently or about 25% of the time

3   Sometimes or about 50% of the time

4   Frequently but not always or about 75% of the time

5   Always, Consistently, or 100% of the time

Step 1 - Self Evaluation

Directions: Read each of the statements below. Think about how often that statement describes your thinking or actions. Using the scoring system above, write the score next to each statement.

Section 1		Section 2	
Statement	Score	Statement	Score
You see contributions as outweighing gain.		You think really BIG consistently.	
You see how everything is connected together like a chain reaction.		You are not afraid to seek help.	
You value other ideas because you are not always right.		You take time out to do nothing.	

Section 1		Section 2	
Statement	Score	Statement	Score
You know ideas take time to develop.		You always have an agenda.	
You see stress as a friend.		You turn tedious tasks into games.	
You always take responsibility.		You listen more than you talk.	
You take time to reflect.		You value experiences over materialistic needs.	
You know how to follow through.		Your ideas have an expiration date.	
You are always learning.		You are always positive.	
You will be humble, and recognition isn't that important.		You ask more questions than answering them.	
You plan ahead but always allow for spontaneity.		You focus on what matters.	
You work to quiet your mind.		You have no limitations.	

Section 1		Section 2	
Statement	Score	Statement	Score
You reject popular thinking.		You make lists but not too many. Completing 3 things a day is a good day.	
You are strategic.		You operate off of the Pareto principle - 80/20 rule.	
You go all in.		You are not afraid to be different.	
Total Section 1		**Total Section 2**	

Section 3		Section 4	
Statement	Score	Statement	Score
You value curiosity over passion.		You consistently bounce ideas off other people.	
You don't just think you do it differently, you do it to test.		You focus your energy.	
You can stay composed.		You value honesty	

Section 3		Section 4	
Statement	Score	Statement	Score
You know to dream big but find ways to remain grounded.		You know that thinking is a discipline.	
You ensure you get exposed to a variety of people.		You are confident.	
You know and do collaborate.		You always know you can do it.	
You value unconventional approaches.		You stay open-minded.	
You are fearless.		You seek out small victories, not large ones.	
You know you will develop your thoughts over time.		You like to learn from your failure.	
No matter the situation, you believe it will work.		You know you are the path and not an obstacle.	
You have a strong, ethical, moral code.		You have grit and determination.	

Section 3		Section 4	
Statement	Score	Statement	Score
You work hard, but it is because you enjoy your work.		You are willing to challenge the status quo consistently.	
You always have an optimistic outlook.		You are consistently persistent.	
You embrace your uniqueness.		You consistently think differently.	
You consistently learn differently.		You consistently work differently.	
Total Section 3		**Total Section 4**	

Step 2 – Total Scores

Take the total from each section above and record it in the table below. Add up all the section to determine your total score.

Total	
Section	**Score**
Section 1	
Section 2	
Section 3	
Section 4	
Total	

Step 3 – Understanding Your Score

There is a maximum score of 300. Take your grand total score from all four columns and compare it to the following:

Score	Meaning
200 - 300	You are in the right frame of mind. You are consistently thinking differently and therefore always ready to fail fast and fail safe.
150 - 200	You have course correction opportunities. Thinking differently is not always possible. Consistently failing fast and safe is forthcoming. The good news is that you have a good mindset and you have the opportunity to keep growing!
Under 150	Thinking differently is something that doesn't come easy for you. You may not be comfortable with failing fast and failing safe, but this is a great opportunity to examine your thinking. You have the greatest potential for developing a new mindset! Don't be discouraged. This is the start of an incredible and exciting journey to failing fast and failing safe.

Step 4 – Creating an Action Plan

Now go back to the items in your list that did not score a 4 or a 5. Circle them. Which ones can you work on to improve that score? Which ones need some extra time to develop habits? Create a plan for a few these to tackle – the ones that you see as important for either yourself or your organization to improve.

About the Self Evaluation Tool

Immediately after completing this self-evaluation tool, inquiring minds want to know how this model was created. Simple. It was formed from the characteristics, skills, and concepts from many of the world's greatest entrepreneurs and thought leaders and what they value in thinking and from failing. Whether you like these people or not, there are lessons that can be learned from Bill Gates, Oprah Winfrey, Steve Jobs, Deepak Chopra, Indra Nooyi, Elon Musk and so many, many more.

Three Lists – Me, My Team, and My Organization

Time to reflect on learning differently. We want you to make three lists. The first list is for you, the second is for your team at work, the third is for your organization. Go back through the entire book and jot down your list of things that you know can make a difference for you, your team and your organization. Is it the ELM model? Is it critical thinking? Start formulating a plan on what you need to do to improve.

Consider the following with each item of interest:

- Who?
    - Who needs to be involved?
    - With who am I working?
    - Who makes the decisions?
    - Who approves?
    - Who accepts failing fast and safe?
    - Who does not accept failing fast and safe?
    - Who is willing to think, learn and work differently?

- Where?
    - Where do we have the opportunity to fail fast and safe?
    - Where will the failure take place?
    - Where should it take place?
    - Where will people be able to think, learn and work differently?

- What?
    - What is our goal?
    - What is our focus?
    - What could be?
    - What will happen if we do?
    - What will happen if we do not?
    - What will people do to think, learn and work differently?

- When?
    - When should we try to fail?
    - When do we start?
    - When do we end?
    - When will people think, learn and work differently?

- Why?
    - Why are we doing this?
    - Why have we not before?
    - Why are trying differently?
    - Why should people think, learn and work differently?

- How?
    - How are we going to approach this?
    - How are we doing it differently?
    - How do we come up with a solution?
    - How is that solution going to operate?
    - How will people react to it?
    - How will people adopt it?
    - How will people think, learn and work differently?

- How Many?
    - How many people are involved?
    - How many processes?
    - How many systems?
    - How many times will we do this?
    - How many more questions do I need to ask?
    - How many people will think, learn and work differently?

These are the 7 fundamental areas that need to be considered when doing practically anything whether it is problem-solving, root cause analysis, charters, self-improvement, failing fast, or failing safe. Using the 7 dimensions model, we ask and answer WWWWWHHM (Who, What, Where, When, Why, How and How Many).

What about working differently? We have one question for you. What are you waiting for?

Let's do one final assessment to see if your organization is ready to work differently.

Organizational Self-Assessment

The self-assessment is designed for organizations to determine how well the organization handles failing fast and failing safe concepts.

Step 1 – Complete Self-Assessment

For each of the statements below, choose from either column A or column B that best describes your organization in relation to the sentence in the statement column. Put the score in the open box between columns A and B.

Statement	Column A		Column B
Example: This happens all the time...	Yes (1 Point)	2	No (2 Point)
The project or effort was designed...	Thoughtfully (1 Point)		Carelessly (2 Points)
The project or effort was...	Collaborative (1 Point)		Resisted (2 Points)
The project or effort had...	Personal Interests Dominated / No Goals (2 Points)		Truthful Goals (1 Point)
Risks, issues, budget, and deadlines were...	Honest (1 Point)		Deceptive (2 Points)

Statement	Column A		Column B
The team made mistakes but…	Mistakes Were Not Repeated (1 Point)		The Same Mistakes Were Made Repeatedly (2 Points)
Failure was _____ addressed or avoided through simple communication or research.	Never (2 Points)		Always (1 Point)
Was experimentation encouraged?	Yes (1 Point)		No (2 Points)
Did anyone play the blame game?	Yes (2 Points)		No (1 Point)
Was the failure detected early?	Yes (1 Point)		No (2 Points)
Is all failure treated equally?	Yes (1 Point)		No (2 Points)
Total Points			

Step 2 – Total Points

Add up all of the scores from the questions and enter the total in the Total Points section provided.

Step 3 – Understanding Your Score

Take your total points and compare it to the following:

Your Score	Score Meaning
10 – 13	Failing fast and safe is acceptable and doing its part to ensure everyone thinks, learns and works differently. Your organization learns from failure and adapts. Your organization is doing awesome!
14 – 16	Your organization is doing some things well and is on the path to becoming an enlightened culture that embraces failing fast and safe. Teams generally understand how to harness failure and learning.
17 – 20	Your organization has some work to do. Teams need to sit down and discuss what it really means to fail fast and safe. Teams need to discuss why it is important to your organization's long-term success.

Final Thoughts on Your Journey to Failing Fast and Failing Safe

"There are no secrets to success. It is the result of preparation, hard work, and learning from failure."

- Colin Powell

What haven't we thought of that everyone needs to know about failing fast and failing safe? A lot! We are sure of that. What we hope is that in our failure you will go on to research, experiment and do all of those things that will allow you to learn and excel. And someday you will pass it on to those you work with, or with us and we will make version 2 of this book someday much stronger. What we won't do is dwell on that failure.

We learn and move forward.

"You build on failure. You use it as a stepping stone. Close the door on the past. You don't try to forget the mistakes, but you don't dwell on it. You don't let it have any of your energy, or any of your time, or any of your space."

- Johnny Cash

Let's be clear that the best people, the best organizations understand failure. James Quincey, President and CEO of Coca-Cola Co. called upon his staff to move beyond the fear of failure they had been living with for years since the "New Coke" disaster. He said, "If we're not making mistakes, we're not trying hard enough."

Build and move on. Learn.

"Failure is simply the opportunity to begin again, this time more intelligently."

- Henry Ford

Netflix co-founder and CEO Reed Hastings went beyond getting over the fear of past failures. He is worried that they have too many hit shows, and they have not been canceling enough shows. He said, "Our hit ratio is too high right now. We have to take more risk… to try more crazy things… we should have a higher cancel rate overall."

What? Really? Yes. With this mindset of experimentation and drive towards failing fast and safe they understand that failing with moderately good shows is great. Experimenting with them will help them find the next big hit.

Experiments pay off.

Amazon CEO Jeff Bezos states clearly that his company's growth is based on its failures. He said, "If you're going to take bold bets, they're going to be experiments. And if they're experiments, you don't know ahead of time they're going to work. Experiments are by their very nature prone to failure. But a few big successes compensate for dozens of dozens of things that didn't work."

Give yourself and your people permission to fail.

Fail Fast – Fail Safe Development Model

The model below helps chart individual or organizational maturity in the concepts of Fail Fast – Fail Safe. Where are you in the model? Where is your organization? Think about how you would move yourself or your organization into the proficient quadrant. What steps would you take to become proficient in failing fast and failing safe?

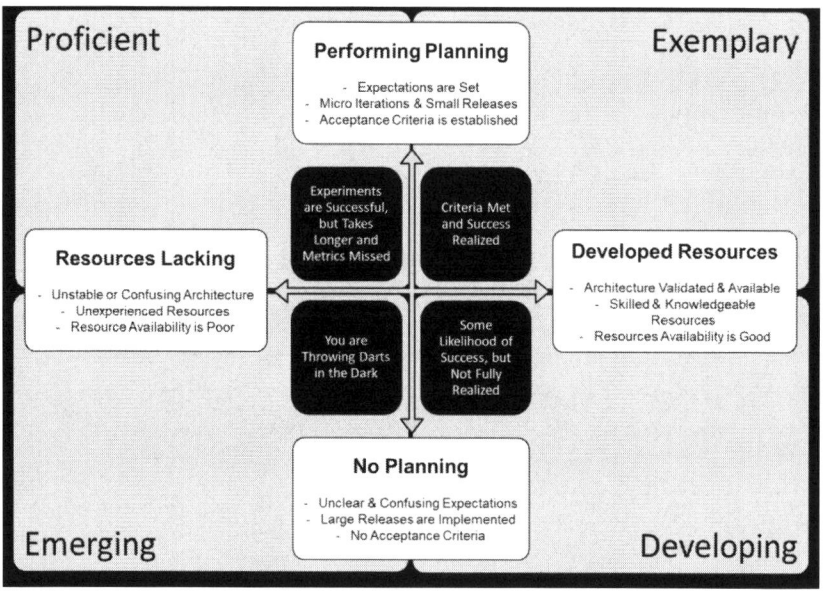

As we cross the finish line on our journey together, we would like to give you a final checklist:

Checklist to Failing Fast and Failing Safe
1. Understand Failure with You and Your Organization.
2. Frame Yourself as a Learner.
3. Become a Change Agent for Failing Fast and Failing Safe.
4. Plan and Work your Model for Failing Fast and Failing Safe.
5. Incorporate Agile into your Failing Fast and Failing Safe.
6. Get some Fun in Your Failure.
7. Be Curious – Always.
8. Don't Fear Failure. Embrace it.
9. Plan and Work your Model for Failing Fast and Failing Safe.
10. Practice Thinking, Learning & Working Differently.

"I can accept failure, everyone fails at something. But I can't accept not trying."

- Michael Jordan

We firmly believe that. You have to try. Paul and Bob have known each other for over 30 years. Failing fast and failing safe has been key in our personal and professional journeys. As we enter our 6th year as an organization of showing organizations how to think, learn, and work differently - we know that failing fast and failing safe is key to success. We thank you for taking this journey of failure with us, and we wish you much success as you fail fast and fail safe!

About the Authors

Bob Prentiss is COO of Bob the BA and Founder of the Uncommon League. Bob is passionate about helping you think, learn and work differently. Collector of all things dragon related and Dr. Who fan. Bob is CBAP certified with 30+ years of experience in corporate America; with a background in managing BA centers of excellence, assessing and managing BA maturity, quality, and competency. Bob is an award-winning curriculum developer. Bob has presented numerous keynote, workshops, seminars, conferences, and training sessions across North America. Bob is a founding member and past President of the IIBA MSP Chapter.

Paul is a Founder of The Uncommon League, CFO & Chief Paper Pusher for The League of Analysts Inc - Bob the BA. Paul was President of Minneapolis-St Paul Chapter which was awarded Chapter of the Year and Most Innovative Chapter by IIBA. Paul has presented at events all over North America including BA World, Project Summit, Local IIBA Chapter Meetings and other events. With over 25 years of experience in manufacturing, insurance, healthcare, education, and government sectors in the roles of CFO, Portfolio Manager, Program Manager, Project Manager, Educator, Editor, Business Analyst and Green Belt Process Improvement, Paul has helped many companies think, learn, and work differently.